AF323983

FINE ARTS, MUSIC AND LITERATURE

MUSIC AND HEARING

Fine Arts, Music and Literature

Additional books in this series can be found on Nova's website
under the Series tab.

Additional e-books in this series can be found on Nova's website
under the e-book tab.

FINE ARTS, MUSIC AND LITERATURE

MUSIC AND HEARING

K. RAJALAKSHMI

nova publishers
New York

Copyright © 2015 by Nova Science Publishers, Inc.

All rights reserved. No part of this book may be reproduced, stored in a retrieval system or transmitted in any form or by any means: electronic, electrostatic, magnetic, tape, mechanical photocopying, recording or otherwise without the written permission of the Publisher.

For permission to use material from this book please contact us:
nova.main@novapublishers.com

NOTICE TO THE READER

The Publisher has taken reasonable care in the preparation of this book, but makes no expressed or implied warranty of any kind and assumes no responsibility for any errors or omissions. No liability is assumed for incidental or consequential damages in connection with or arising out of information contained in this book. The Publisher shall not be liable for any special, consequential, or exemplary damages resulting, in whole or in part, from the readers' use of, or reliance upon, this material. Any parts of this book based on government reports are so indicated and copyright is claimed for those parts to the extent applicable to compilations of such works.

Independent verification should be sought for any data, advice or recommendations contained in this book. In addition, no responsibility is assumed by the publisher for any injury and/or damage to persons or property arising from any methods, products, instructions, ideas or otherwise contained in this publication.

This publication is designed to provide accurate and authoritative information with regard to the subject matter covered herein. It is sold with the clear understanding that the Publisher is not engaged in rendering legal or any other professional services. If legal or any other expert assistance is required, the services of a competent person should be sought. FROM A DECLARATION OF PARTICIPANTS JOINTLY ADOPTED BY A COMMITTEE OF THE AMERICAN BAR ASSOCIATION AND A COMMITTEE OF PUBLISHERS.

Additional color graphics may be available in the e-book version of this book.

Library of Congress Cataloging-in-Publication Data

Music and hearing / K. Rajalakshmi (Department of Audiology, All India Institute of Speech & Hearing, Manasagangothri, Mysore, Karnataka, India).
pages cm. -- (Fine arts, music and literature)
Includes index.
 ISBN 978-1-63463-621-6 (hardcover)
 1. Musical perception. 2. Hearing. 3. Music--Acoustics and physics. I. Rajalakshmi, K., 1961-
ML3838.M94945 2015
 781.1--dc23
 2014045011

Published by Nova Science Publishers, Inc. † *New York*

"Across galaxies of time and space
Travelling just to see your face
Lost amidst the countless stars
To bring me back to where you are."
This is for you, Dear SON
this is for your love, mine is yours
Love is fate, I am here
Because you know the meaning of life
That begins and ends with a kiss/Love/Breath
We are knights in shining ardor, who toil for you
And our children, it's a circle
So they will know this truth
Love is the sacred gospel, all we need to know
As your mother and lover, my spirit lives imbued
With, from and by your wisdom and beauty
I am here to pay honor and homage to your soul
This is and will always be my devotion
This I dedicate, because through you, I become whole"

Dedicated To My Dear Son Anil Jagadish Who Was Unconditional In Loving,
Always Undearstanding, A Helping Friend And Soulmate, Who Left Me
Amidst This World Of Confusion, Sorrow And Misery As He Was Summoned
By The Supreme Power For A Journey Of Greater Accomplishments.
Love you dear. You are always my inspiration and purpose for life.

CONTENTS

FOREWORD

Foreword from Jagadguru Sri Sri Sri Jayendra Puri Mahaswamiji

Shreemukha of H.H. Parama Pujya Srimatparamahamsa Parivrajakacharya Sachidananda Pranavaswarupa Acharya Mahamandaleshwar Jagadguru Sri Sri Sri Jayendra Puri Mahaswamiji, Peethadhipathi of Sri Kailash Ashrama Mahasamsthana, Rajarajeshwarinagar, Bengaluru, 560 098.

Shishurvetti pashurvetti vetti gaanarasam phanih

The bliss of music is known to a child, an animal and also a serpent.

This is the glory ascribed to music which is unaffected by any boundaries whether geographical, language or any other social criteria. There is no limit to the endearing capacity of music. It can melt stones, produce rain, drive away darkness, turn the wicked into virtuous, heal illnesses, or induce sleep in a child or adult. It can pacify animals, hypnotize serpents, befriend cattle, enhance the progress of plant growth and so on and so forth. It is such wonderful talents that even in the absence of any other person, music can soothen or ignite once own heart when sung in isolation. It can force divine forces to descend to this mortal world and yearn to be mortal too. It can bind, release, imprison, free, harden, soften, inspire, discourage, encourage; there is almost nothing that music cannot do. Music is pleasing even when it is regulated by strict rules or even when it is spontaneously rendered, filled with deep sentiments.

Blessed are those that are endowed with this talent naturally and also those who acquire this talent by systematic study. Music, as a performing art, especially for those who have adopted it as their profession, has immense value for their living and for the contribution of social joy through entertainment. They would have to certainly protect the means of such capacities which are basically the vocal chords and the auditory functions. Although music has grown as naturally as the art of drawing and painting, it has a special significance as a performing art and it is almost natural that it has evolved into several systems over a period of time. It is through these systems that it can be well acquired as a formal academic qualification even by those who were not born with such a talent.

This book compiled by Dr. Rajalakshmi K on this subject is bound to draw the attention of both naturally talented professional musicians and others who have acquired it through a systemetic study and aid them immensely to protect, enhance or to sharpen their talents simultaneously maintaining their overall health and longevity. Her effort is truly laudable and deserves all appreciation.

We pray to Divine Mother Sri Jnanakshi Rajarajeshwari to grant all readers a deeper insight into this aspect of music and a long and fruitful future to Dr. Rajalakshmi K and all others who have contributed in the completion of this book.

SRI DEVI SMRITIH

In the service of Divine Mother and Mahaswamiji
K. Singaravelu.

ACKNOWLEDGMENTS

This book could not be written to its fullest without the grace of DIVINE MOTHER, who served as my guide as well as one who challenged and encouraged me throughout my time spent studying. She would have never accepted anything less than my best efforts, and for that, I thank MOTHER.

To My Mother,
Without her none of this would be possible.

To My Father
A very loving person who taught the values and joy of sharing and caring. For the freedom he provided us and instilling greater self confidence in us.

To My Husband,
For his constant love and support, and remembering to take care of me when I couldn't remember to do so myself.

To my siblings
Perhaps the most faithful companions I've had. I love them all. I would like to thank the rest of my family, who supported and encouraged me in spite of all the time it took me away from them. It was a long and difficult journey for them.

To my SAATHIS,
The light of my life, my rock, my safe place.

I thank my Director and senior Dr. S. R. Savithri for her inspiration and encouragement.

I would extend my heartfelt thanks to all my teachers in school and professors in College who moulded me to the person I am today.

My thanks are due to my Postgraduate students Anoop O.T, Abhishek Saha, Arya Chand, Deepika Verma and Zubin Vinod who worked on their master's dissertation and Priyanka who is working at present on her doctoral thesis in this subject. I also thank my research project staff Apoorva, Bharath, Anoop and Merin.

My pillars of support and springs of love Kripal and Vinaya Manchaiah.

I would like to express my gratitude to my people who saw me through this book; to all those who provided support, talked things over, read, wrote, offered comments, allowed me to quote their remarks and assisted in the editing, proof reading and design. Apoorva, Vinaya Manchaiah, Rangamani, Priyanka, Zubin, Dhananjay and Rama.

I would like to thank Spoorthi T., for helping me in the process of selection and editing.

Thanks to Nova Science, my publisher who encouraged me.

I wish to thank everyone who helped me complete this book. Without their continued efforts and support, I would have not been able to bring my work to a successful completion.

Last and not least: I beg forgiveness of all those who have been with me over the course of the years and whose names I have failed to mention.

Chapter 1

INTRODUCTION TO MUSIC

1.1. WESTERN MUSIC

The origins of "Western music" lie in the early middle Ages and the plainsong of the Church of Rome. Until the end of the middle Ages, the development and spread of music was inherently linked to religious ceremonies. Singing was used primarily during the rites of Holy Communion and other ceremonial gatherings of the Church. In such circumstances music was as simple and pure as possible, justifying the choice of monody (a single voice line, as opposed to polyphony) and the absence of a defined rhythm. The measured rhythm first appeared in the thirteenth century. This pared-down, self-sufficient liturgical music became enriched through the efforts of St. Ambrose, the fourth-century bishop of Milan. He discovered eastern modes, in particular the hymns of the Byzantine Church with a scale structure borrowed from Greek modes, where the intervals between notes were smaller. Ambrose used these modes in the liturgical music of the mass. The singing became more fluid and sinuous. Soon he would leave behind even the classical Greek modes.

The Ambrosian chant spread throughout Europe so successfully that it was adopted and altered in faraway lands, where it came under the influence of regional languages and musical modes. The Church, anxious to preserve the pure and universal nature of its music, wished to prevent this. Writing notes down on paper occurred in the seventh century, when various signs were used to codify how to sing in church. At the beginning, notation was simply an aid to memory. Then it developed into proportional notation, with a four-line staff and square or diamond-shaped notes.

The cadence as a basic element makes it possible to study forms after this period. Leo Stein defined it as a moment of rest marking the end of a phrase or section, noting that the word "cadence" was derived from the Latin *cadere* (to fall), since a feeling of *caesura* or pause was implicit in the sound of a low note followed immediately by a higher one. This analogy between spoken and sung phrases is important because it provides a key to formal construction.

Religious music within the confines of the medieval Church did not develop much, since religious authorities were concerned with preserving it unchanged. Outside the Church, however, it exerted great influence on secular music. Cadences was used to give rhythm to religious singing and provide reference points. In popular songs, they served as a form of punctuation and organization. Songs with very simple melodies and rhythms were made up of short couplets with differing texts. The invention of the bar, attributed to troubadours and *trouvères* of France, wandering minstrel-poets of the north and south respectively, provided rhythmic unity and precision. The use of the bar gradually ousted plainsong and had a tremendous influence on written music.

Now, music was on the way to becoming more complex and a source of pleasure for all. In *The Evolution of Musical Form* (London: Oxford University Press, 1943), Edward C. Bairstow notes certain pivotal occurrences in the growth of form. He points out that works of art display characteristics that can be divided into two categories: those that express unity and those that provide variety. He adds that unifying elements are common to all works, and some, like rhythmic patterns, have been common to the music of all time. Rhythmic patterns are more easily memorized by ear, so it's no surprise to learn that they can also be used to differentiate melodic themes that, through contrast and repetition, are the basis for developing various parts of a work.

The best known musical form is certainly the ABA, where A designates the principal section and B, a contrasting section in another key. Thousands of arias in operas, oratorios, and cantatas have used this form, as have many instrumental pieces. The minor ternary (ABA) form is the precursor of such dance forms as the minuet, march, waltz, or scherzo, which have the same tripartite organization but provide contrast through rhythm, the number of *reprises*, changes in harmony, and so forth.

The influence of these minor dance forms is far from negligible. They are the forerunners of major forms such as the sonata, or minor forms enlarged through repetition. They have provided basic material for many composers, not to mention analysts and theoreticians who are forever giving new interpretations to the monuments of music.

1.2. Indian Classical Music

Indian classical music is an art form of the Indian Subcontinent. The origins of Indian classical music can be found in the Vedas, which are the oldest scriptures in the Hindu tradition. The Samaveda is derived from the Rig-Veda so that its hymns could be sung as Samagana. This chanting style evolved into jatis and eventually into ragas. Indian classical music has also been significantly influenced by Indian folk music. Bharat's Natyashastra was the first work to lay down fundamental principles of dance, music, and drama.

Indian classical music is both elaborate and expressive. Like Western classical music, it divides the octave into 12semitonesof which the 7 basic notes are, in ascending tonal order, *Sa Re Ga Ma Pa Dha Ni* for Hindustani music and *Sa Ri Ga Ma Pa Da Ni* for Carnatic music, similar to Western music's *Do Re Mi Fa So La Ti*. Achal Swaras are the fixed swaras of the seven musical notes. Sa and Pa are the achal swaras of the Indian classical music. The term Arohi, also known as Arohana and Aroh, is used to define the ascending melody in music. However, Indian music uses just-into nation tuning, unlike most modern Western classical music, which uses the equal-temperament tuning system. Also, unlike modern Western classical music, Indian classical music places great emphasis on improvisation. Indian classical music is monophonic in nature and based around a single melody line, which is played over a fixed drone. The performance is based melodically on particular ragas and rhythmically on talas.

Music has always been an important part of Indian life. The range of musical phenomenon in India extends from simple melodies to what is arguably one of the most well- developed "systems" of classical music in the world. The great poet-saints Surdas, Tulsidas, who chose to communicate in vernacular tongues brought forth a great upheaval in north India and the Bhakti or devotional movements they led by Kabir and Mirabai gained many followers and continue to be immensely popular. By the sixteenth century, the division between North Indian (Hindustani) and South Indian (Carnatic) music was also being more sharply delineated. Classical music, both Hindustani and Carnatic, may be either instrumental or vocal.

Origins of Indian Music

Indian music is probably the most complex musical system in the world with a very highly developed melodic and rhythmic structure. The structure includes complicated poly-rhythms, delicate nuances, ornamentations and microtones which are essential characteristics of Indian music. This makes it very difficult to notate every detail in Indian music.

Originally Indian music was passed on by oral tradition (*Guru:* teacher; *shishya*: student and *parampara*: tradition) from one generation to another for centuries. The music was never written down until much later.

The notation system was actually developed much later more as a memory-aid than something from which to learn or something from which to perform. This is why the tradition wherein the student learns from a *Guru*on a "one-on-one" basis is considered to be the only real way to learn music since there are so many aspects that cannot be learned from a book because the existing notations are only a skeletal representation of the music.

The ancient period originates with the writing of the *Vedas* and is followed by the *Upanishads*. Though the *Upanishads* are considered by some as the concluding portion of the *Vedas* or the *Vedanta*, they are in a class by themselves. Dating from 1000 - 300 B.C., they laid the foundations on which the philosophies and religions of India are based. It is in the *Upanishads* that the solfege system of seven notes is discussed. In the West, the solfege system was not developed until the tenth century A.D. by Guido of Arezzo. The period of the epics, the Ramayana and the Mahabharata (500 B.C. - 200 A.D.) saw the development of the Jati system on which the modern Raga system is based. Also, various melodic and percussion instruments are mentioned during this time. Mention must be made of the *Natyashastra*, a treatise on the performing arts written by Bharata in 300 B.C. It is the most authoritative and ancient work on the classical science of music and dance.

Another milestone in the development of Indian music is the *Brihadesi* whose author, Matanga, started a scientific classification of scales which was the basis for the later development of the seventy two *Melakarta* system (parent scales) by the great scholar Venkatamakhi. Narada, another contemporary scholar of Matanga, further defined Ragas, coded the microtones (*Srutis*) and gave them their twenty-two names.

The medieval period dates from approximately the fifth to the seventeenth centuries A.D. The *Geeta Govinda* and the Indian *Song of Songs* were composed during this period. The medieval poet Jayadeva was the first to introduce the concept of *Chhanda Prabandha* (verses set to rhythm in a uniform manner.) The text, in Sanskrit, consists of beautiful songs dealing with the *Radha Krishna* theme - thus, it is of religious and musical importance. Jayadeva's songs, known as *Ashtapadi* are still sung today though the melodies may vary.

The *Sangeeta Ratnakara,* another treatise by the great composer and musician Sarangadeva, deals with the classification of Ragas according to the various seasons and different times of the day and the importance of certain notes in the delineation of the Raga (*Vadi* and *Samvadi*)

During the twelfth and thirteenth centuries, northern India endured a series of invasions by Muslim rulers from Asia Minor on a crusade to spread the Islamic religion throughout the region.

Until this time, the history of Indian music remained the throughout the whole of India. Following the Islamic invasions and the concurrent cultural amalgamation, Indian music developed two distinctive systems: North Indian music (*Hindustani music)* and South Indian music (*Karnatic or Carnatic music.*) The influence of Islam together with other cultural, social and political forces produced the unique *Hindustani* (literally, the music of India – *Hindustan:* the Hindi word for India, *Stan:* land) style. In South India music continued to develop without external influence and is still known today as *Karnatic* (literally - in Tamil - old or traditional).

The modern period (seventeenth century A.D. to the present) unfolds as a result of the efforts of individuals. Other than Venkatamakhi, the scholar who devised the system of seventy two *Melakartas* (parent scales from which thousands of *Ragas* are derived), three composers, known today as the musical trinity, worked and flourished, (*Tyagaraja-* 1767-1847; *Muttuswamy Dikshitar-* 1776-1835 and *Syama Sastri-* 1762-1827). Born in *Tiruvarur,* Tamil Nadu, all three were scholars of the scriptures and sacred literature.

Tyagaraja (1767-1847)

Figure 1.2.1.Tyagaraja.

Born on May 4, 1767, *Tyagaraja* was a prolific composer who composed more than 2,000 songs. His compositions were mainly in the Telugu language though he did compose in *Sanskrit* too. He wrote two Operas, *Prahlada Bhakti Vijayam* and *Nowka Charitram*. A great devotee of Lord Rama, Tyagaraja's songs are sublime and soul-stirring. He was the first composer to perfect the musical form *Kriti*. He also introduced the concept of the *Sangati* (variations on the melodic line of a composition which can enrich the composition. He placed great importance and emphasis on the value of absolute music. He wrote several songs in unusual *Apurva Ragas*. His style is simple, beautiful and charming which appeals not only to the scholar but also to the layman. He also composed songs in some of the lesser used *Melakarta Ragas*. His *Pancharatna kritis* (*pancha*: five & *ratna*: jewel or gem) are a class by themselves. They are based on the five *Ghana Ragas*: *Natai, Gaulai, Arabhi, Varali* and *Sri*. They are unusual in form and consist of a *Pallavi, Anupallavi* and multiple *Charanams*. The *Charanams* are first sung in solfa syllables and then repeated with the text.

Muthuswamy Dikshitar (1776-1827)

Figure 1.2.2. Muthuswamy Dikshitar.

Born on March 24, 1776, *Dikshitar* was the youngest of the Trinity. He wrote about 300 compositions in the *Sanskrit* language. A great scholar, he had profound knowledge of the *Vedas, Upanishads*, astrology, mythology, magic, etc. He sang praises of all the Deities without exception. He chose a medium to slow tempo for his songs which gave him the scope to bring out the depth and beauty of each *Raga* by using subtle *gamakas* (ornamentations) and delicate microtones. The use of the *Madhyamakala* (passages which have a faster tempo than the rest of the song) only added to the beauty of the compositions. His great intellect shows in all his compositions. He has compositions in many rare ragas and talas. His five year stay in the holy city of Benares caused him to be profoundly influenced by the *Dhrupad* style of singing which was prevalent at the time. He also composed songs based on some North Indian *Ragas*. His *Navagraha kritis* in praise of the nine planets reveal his knowledge of the science of astrology. These were originally seven in number called the *Vara kritis* after the seven days of the week and were based on the basic seven *Talas*. The two *kritis* in praise of *Rahu* and *Ketu* were added later. The *Navavarana kritis* in praise of Devi (Goddess) contain some of his other superb compositions.

Syama Sastri (1762-1827)

Figure 1.2.3. Syama Sastri.

Syama Sastri was born on April 26th, 1762. He was well-versed in both *Telugu* and *Sanskrit*, both of which he used in his compositions. He composed about three hundred works, mainly *kritis* and *Swarajatis*. Being a great devotee of Devi (the goddess Parvati), she is the theme for his compositions. He perfected the *Swarajati* to its present form. His *Swarajati* in the *Raga Todi* is a beautiful example of this. Some of his compositions are very rich in rhythmic conception. His style is not as simple as Tyagaraja's but at the same time not as laborious as Dikshitar's. Here below are the key notes to remember history of Indian music:

- Ancient Period (4000 B.C. - 400 A.D.)
  Vedas (4000 B.C. - 400 A.D.) Upanishads (1000 B.C. to 300 A.D.)Ramayana and Mahabharata (400 B.C.) Bharata's Natyasastra (300 B.C.) Matanga's Brihadesi (400 A.D.) Narada (400 A.D.)
- Medieval Period (400 A.D. - 1600 A.D.)
  Jayadeva's Geeta Govinda (12th Century) Sarangadeva's Sangeeta Ratnakara Bifurcation into two systems of music Purandara Dasa (1484-1564)
- Modern Period (1600 A.D. - present)
  Venkatamakhi (early 17th Century) Shyama Sastri (1762-1827) Tyagaraja (1767-1847) Muthuswamy Dikshitar (1776-1827)

Carnatic Music

Carnatic music or Carnatic Sangeet is the classical music form of south India. Carnatic music has a rich history and tradition and is one of the gems of world music. Carnatic Sangeet has developed in the south Indian states of Tamil Nadu, Kerala, Andhra Pradesh and Karnataka. These states are known for their strong presentation of Dravidian culture. Purandardas (1480-1564) is considered to be the father of Carnatic music. To him goes the credit of codification of the method of Carnatic music. He is also credited with creation of several thousand songs. Another great name associated with Carnatic music is that of Venkat Mukhi Swami. He is regarded as the grand theorist of Carnatic music. He also developed "Melankara", the system for classifying south Indian ragas.

It was in the 18th century that Carnatic music acquired its present form. This was the period that saw the "trinity" of Carnatic music; Thyagaraja, Shamashastri and Muthuswami Dikshitar compile their famous compositions. Numerous other musicians and composers have also enriched the tradition of Carnatic music. Some other notable Carnatic music exponents are Papanasam Shivan, Gopala Krishna Bharati, Swati Tirunal, Mysore Vasudevachar, Narayan Tirtha, Uttukadu Venkatasubbair, Arunagiri Nathar and Annamacharya.

In Carnatic music there is a very highly developed theoretical system. It is based upon a complex system of Ragam (Raga) and Thalam (Tala). Raga is basically the scale consisting of the seven notes Sa, Re, Ga, Ma, Pa, Dha and Ni. However, unlike a simple scale, there are definite melodic restrictions and compulsions. The Ragams are classified into various modes. These modes are referred to as mela, which are 72 in number. The Tala (thalam) is the rhythmic foundation of the Carnatic music.

It is by no means an easy task to define music but one may understand it as a special means of communication through organised, regular vibrations. Audible frequencies can be classified on the basis of regularity of vibrations into noise, sound and music. We *hear* noise, *listen* to sound and *enjoy* music. Like many other systems across the world, Carnatic music can be appreciated mainly at two levels - emotional or intellectual. It has generally been observed that the majority, at least in the initial stages, go by 'what it does to their heart'. But gradually they seek to understand more in order to enjoy the music better. Of course, there are many who enjoy it for spiritual, philosophical and other reasons.

Carnatic music is one of the two major systems of classical music in India, the other being *Hindustani music*. One of its greatest virtues is that while it is among the most scientifically evolved independent systems in the world, it has also managed to take in desirable aspects from other systems and adopt them with an enviable catholicity of outlook, without in any way prejudicing its originality and individuality. For instance, the violin has been successfully adopted from the West, just as a few ragas have been incorporated from Hindustani music.

Both Hindustani and Carnatic music use the system of ragas—sets of pitches and small motives for melody construction—and tala for rhythm. Ragas form a set of rules and patterns around which a musician can create his or her unique performance. Likewise, tala is a system of rhythmic structures based on the combination of stressed and unstressed beats. Within these rhythmic structures, musicians can create their own rhythmic patterns building off the compositional styles of others.

Musical Forms

Musical forms may be categorized under two headings: 1. Abhyasa Gana (literally abhyasa - practice; gana - song) 2. Sabha Gana (literally sabha - audience). These forms, therefore, (1) intended for practice or to acquire technical skills and/or (2) intended for performances before an audience in a concert situation. Originally, all compositions were composed as vocal music and the same compositions are played on various instruments. Therefore, all forms have texts.

Gita or Gitam

Literally, "song", Gitam belongs to the Abhyasa Gana group and denotes a particular type of composition. It is a very simple musical "paragraph" that has no divisions. It is sung in a uniform tempo from the beginning to the end without variations. The melody is very simple and outlines the Raga on which it is based and is set to a particular Tala.

Swarajati

Swarajatis, which also belong to the Abhyasa Gana or technical group, are compositions which are learned following Gitams. They prepare "the way" for the more complicated form, Varnam. They are similar in structure to the Varnam. Divided into three sections called Pallavi, Anupallavi and Charanam respectively, the Charanams have different tunes. It was originally a dance form containing Jatis (rhythmic syllables). These were later excluded by Syama Sastri who developed and perfected the Swarajati into its existing form today.

Jatiswaram

While in the Swarajati the stress is on the music or the melody, in the Jatiswaram the rhythmic patterns dominate. Very similar to the Swarajati, the Jatiswaram belongs to the repertoire of dance music. Originally the Pallavi, Anupallavi and part of the Charanam were intended to be sung in Jatis (rhythmic syllables). Gradually this practice changed and the Jatiswaram is now sung entirely in Swaras (solfa syllables).

Varnas or Varnams

Though it belongs to the technical group Abhyasa Gana, the Varnam is also performed in concerts as opposed to the Gitam which is not. Literally meaning "colour," the Varnam consists of two halves: the Purvanga or the first half and the Uttaranga, or the second half. The two halves are almost equal in length. The first half consists of the Pallavi, the Anupallavi and the Muktayi or Chitta Swara's. The second half consists of the Charanam and the Charanam Swaras.

The Pallavi and the Anupallavi are usually two lines each, with both sections consisting of lyrics. The Chitta Swaras is a passage of solfa syllables. After the first section is completed, the performer goes back to the first line of the Pallavi. If the Varnam is to be performed in more than one tempo, the first section is played or sung in all the different tempi before proceeding to the second section.

The Charanam consist of one line with text. The Charanam Swaras are groups of solfa syllables. There may be four or more groups of Charanam Swaras in each Varnam. The Charanam is used as a theme of return after each group of Swaras. As in the first half, each group of Swaras is played in all the different tempi before going on to the next group.

Kriti

A Kriti is a composed composition set to a certain Raga and a fixed tala. There are three sections in a Kriti: Pallavi, Anupallavi, and Charanam.

Ragam, Tanam, Pallavi

Ragam

Ragam consists of free improvisation (without rhythmic accompaniment) based on a particular Raga. The soloist develops the Raga in stages, staying within the framework of the Raga. There are certain rules which must be observed and some restrictions that apply. Each Raga is based on a scale of five, six or seven notes.

There are certain notes in the Raga which are more important than the other notes. These are called Vadi and Samvadi and are stressed more than the others during the improvisations. The soloist will not use notes that are not in the Raga (vivadi swaras). If there are any microtones incorporated with any of the notes, they must be used. There are certain typical phrases or usages of certain phrases in some Ragas which make them easily distinguishable. Ragas are derived from Melakartas or parent scales. The Raga Alapana (delineation of the Raga) starts slowly bringing out the beauty and mood of the Raga and is slowly built up, eventually ending with Pharans (fast runs) where the performer can demonstrate his virtuosity and technical prowess.

Tanam

Tanam is the second phase of the Ragam, Tanam and Pallavi where the performer continues to improvise, still without any rhythmic accompaniment. Though there is no drum accompaniment, this section introduces an element of rhythmic pulse as opposed to the Ragam wherein the improvisation is free. At the end of each phrase section, a stereotypical rhythmic cadence pattern is used to indicate the end of that particular section.

Pallavi

Pallavi consists of a short precomposed melodic theme, with words, which is usually set to one cycle of Tala. The theme is played two or three times in its simple form (without variations) during which the drummer familiarizes himself with it and enters. The Pallavi has the following main features:

1. Neraval: This literally means filling up or spreading; in other words, filling up portions of the Pallavi line with new, fresh and creative music. The soloist improvises new melodies built around the words of the Pallavi keeping the rhythmic structure constant.
2. Tri-Kalam: In this section the Pallavi is played in three tempo keeping the Tala or rhythmic cycle constant: i.e. (1) usually twice as slow as the original tempo, (2) the original tempo and (3) twice as fast as the original tempo.
3. Swara Kalpana: This improvised section is performed using swaras (solfa syllables) in medium and fast speeds. Each swara kalpana passage returns to the beginning of the Pallavi theme. The possibilities are endless in this type of improvisation and are only limited by the creative capacity, technical and musical abilities of the individual performer.

Ragamalika: The Pallavi usually ends with this section, which literally means, "Garland of Ragas." The soloist improvises freely in different Ragas and at the end of each Raga comes back to the rhythmic theme of the original Pallavi.

Hindustani Music

During the twelfth and thirteenth centuries northern India endured a series of invasions by Muslim rulers from Asia Minor on a crusade to spread the Islamic religion throughout the region. Until this time, the history of Indian music remained the same for the whole of India. Following the Islamic invasions and the concurrent cultural amalgamation, Indian music developed two distinctive systems: North Indian music (*Hindustani music*) and South Indian music (*Karnatic or Carnatic music.*) The influence of Islam together with other cultural, social and political forces produced the unique *Hindustani* style.

In the period from 600CE to 1200CE came the development of regional music in India and the formation of Hindustani music. Until the 13th century, India had one distinctive type of music. This changed with the invasion of the Mughals. When the Mughals began to live in India, they brought with them outside musical influences from the Arabian and Persian cultures. Music of north and central India began to blend with new styles and created a new style of music called *Hindustani* music. Hindustani music is composed using seven basic notes which are called the ragas. This style was more for entertainment rather than for spiritual purposes. However, music in the south of India was able to preserve the music of their ancestors and continues to be classical *Carnatic* music to this day.

The period from 1200CE to 1700CE was marked by an Indian composer by the name of Amir Khusro, 1253-1325CE. He wrote music in many different languages and invented many different forms of music which include Qawali, Qasida, Naqsh, and Qalbana. He created a new music theory called *Indraprastha Mata* and introduced different music genre called T*arana* and K*aul*. In addition to this, he was the inventor of the sitar and the tabla. He also introduced secular music.

Figure 1.2.4. Amir Khusrow, iconic figure in the cultural history of India.

One of the main differences between North Indian and South Indian music is the increased influence of Persian music and musical instruments in the North. From the late twelfth century through the rise of British occupation, North India was under the control of a Muslim minority that was never able to extend its sphere of influence to South India. During this time, the music of North India began to acquire and adapt to the presence of Persian language, music, and musical instruments, such as the setar, from which the sitar got its name; the kemancheh and santur, which became popular in Kashmir; and the rabab [alternately known as rebab and rubab], which preceded the sarod.

New instruments were introduced, including the tabla and sitar, which soon became the most famous Indian musical instruments worldwide. Legend has that the tabla was formed by splitting a pakhavaj drum in half, with the larger side becoming the bayan and the smaller side the dahini. The barrel-shaped pakhavaj drum, which was the ancestor of both the tabla and the mridangam, has been depicted in countless paintings and prints. New genres of music were formed as well, such as Khyal and Qawwali that combine elements of both Hindu and Muslim musical practice.

Raja Mansingh (1486-1516) introduced the Dhrupad genre of Hindustani music that continues to be enjoyed today. He changed songs written in Sanskrit to Hindi and composed pieces not only for worshipping the Hindu gods, but songs that would relate to the daily lives of people as well.

Mian Tansen, 1506-1589 CE, was a remarkable composer in Hindustani music. He minimized the 4000 ragas and raginis into an arrangement of 400. He also minimized the 92 talas of his time into 12. During the Mughal Era, 16th to 18th century, traditional temple music was replaced with a new style of music called Darbar Sangeet. Akbar, the ruler at this time, enjoyed this style of music as it was used to praise others. He had many musicians in his court that varied from Hindus to Iranis to men and women. The music was very rich and cultured and is proof of the coalition of Indian and Persian music.

Ashtachap is a music period that was created in the early 17th century. It was an older style of music that was used in a religious cult. The cult has created a longstanding temple-based music called the "Haveli Sangeet" After this period, came the Pushti-sangeet period. Hindustani music further evolved in this time as music started to separate from dance. Dhrupad, Khayal and Tappa began. Since 1700CE, the music in India has been continually changing to what is known as Hindustani music today. In this modern period came two new forms of music: Khayal and Thumri. The Khayal genre was formed by Nyamatkhan under the rule of one of the many emperors following the decline of the Mughal Empire. Thumri is a romantic and devotional sort of music that developed in the 19th century.

Figure 1.1.5. Mian Tansen, a depiction by Lala Dea Lal, displayed in Calcutta Gallery.

Another genre of music that was established in this era was the Bengali tappa. It was created by Ramnidhi Gupta (1741-1839CE) and is secular. Pandit Vishnu Digambar Paluskar and Pandit Vishnu Narayana Bhatkhande (1872-1931CE) were the first to introduce post secondary schooling for music. This in turn raised the social standings of musicians.

Hindustani Gharanas

The term Gharana in Indian Classical music means a school or a style of music. It implies a family and a Guru- Shishya Parampara (teacher-student tradition) wherein the style of rendition of the student was a typical trait of its gharana itself. Traditionally from the past all the maestros of a Gharana have been very possessive of their Gharanas and have always tried to ensure that their Gharana maintains its own distinct entity.

There have been Gharanas in Hindustani Vocal, Tabla, Sarod etc. The Gharanas in Hindustani Vocal include Kirana, Patiala, Benaras, Rampur, Sahaswan, Delhi, Indore, Maihar, Gwalior, Agra, Jaipur-Mewati, Jaipur-Atrauli, Bhendi Bazaar and others to name a few. There is a rich tradition of Gharanas in classical Hindustani music. The music Gharanas are also called styles. These schools or Gharanas have their basis in the traditional mode of musical training and education. Every Gharana has its own distinct features. Hindustani classical music is an Indian classical music tradition. It originated in North India around the 13th and 14th centuries. In contrast to Carnatic music, the other main Indian classical music tradition from South India, the

Hindustani classical music was not only influenced by ancient Hindu musical traditions and Vedic philosophy but also by the Persian elements.

Hindustani classical music is known largely for its instrumentalists, while Carnatic classical music is renowned for its virtuosic singing practices. Instruments most commonly used in Hindustani classical music are the sitar, sarod, tambura, sahnai, sarangi, and tabla; while instruments commonly used in Carnatic classical music include the veena, mridangam, kanjira, and violin. The use of bamboo flutes, such as the murali, is common to both traditions as well as many other genres of Indian music. In fact, many of these instruments are often used in both North and South India, and there are many clear relationships between the instruments of both regions. Furthermore, often instruments that are slightly different in construction will be identified by the same name in both the south and the north, though they might be used differently.

1.3. INSTRUMENTAL MUSIC

A musical instrument is an instrument created or adapted to make musical sounds. In principle, any object that produces sound can be a musical instrument—it is after being given purpose that the object becomes a musical instrument. The history of musical instruments dates to the beginnings of human culture. Early musical instruments may have been used for ritual, such as a trumpet to signal success on the hunt, or a drum in a religious ceremony. Cultures eventually developed composition and performance of melodies for entertainment. Musical instruments evolved in step with changing applications.

The date and origin of the first device considered a musical instrument is disputed. The oldest object that some scholars refer to as a musical instrument, a simple flute, dates back as far as 67,000 years. Some consensus dates early flutes to about 37,000 years ago. However, most historians believe that determining a specific time of musical instrument invention is impossible due to the subjectivity of the definition and the relative instability of materials used to make them. Many early musical instruments were made from animal skins, bone, wood, and other non-durable materials.

Music is found everywhere in this beautiful world. As every art needs a strong medium to express its apparatus and uniqueness, similarly, a musical instrument is a medium to express the musical 'Nada' (*lit.* sound), which has significance all over the world. Instrumental Music has an important place in

Indian Music. It is one of the threefold aspects of Music (Vocal Music, Instrumental Music, and Dance) also known as 'Sangeet' in Indian Music. The Instrumental Music is known as "Vadhya Sangeet"

From the beginning of Instrumental Music, two components are very important-the Instrument and the Instrumentalist or Instrument player (Artist). Music is known as a direct way to convey feelings from ancient times. Vocal music consists of words of any language and musical notes which are also known as 'Swaras'. In instrumental music, Swaras are same but boles are played on the musical notes to produce music by Musical Instruments. For instance, in the compositions of plucked instruments, boles are played such as: Da, Dir, Dara etc. Indian Classical Music, as explained earlier, is primarily individualistic. Our instruments are also, therefore, designed to be played solo. Indian Musician considers music as a medium to reach divinity and he therefore is engaged in swarasadhana (tone-culture) while singing or practicing instrument. He is so engrossed in his practice that he becomes completely obvious of his surroundings and his personality merges in a meditative state 'Nadabrahma'. The solo character of our musical instruments is in line with purely individualistic approach. Since there are no fixed or pre-written compositions in Indian Classical Music, every instrumentalist is a composer himself and has absolute freedom to handle the medium of his choice within the boundaries laid down by the exposition of a raga. Attempts have been made in recent times by eminent musicians to use Indian instruments in a group and produce an orchestra effect in the manner of orchestral musical composition in the West, but these have not been very effective. Most of the instruments sound weak and ineffective in a group, since by conception and design, they are meant to be played only in solo and, therefore, are not capable of creating the desired musical impact. There is a traditional system for the classification of instruments. This system is based upon; non-membranous percussion (*ghan*), membranous percussion (*avanaddh*), windblown (*sushir*), plucked string (*tat*), bowed string (*vitat*). In addition to these traditional five classes we have been forced to create a sixth class to accommodate purely electronic instruments. Here below is a list of Instruments (with a brief description) commonly used in India.

Mridangam

The mridangam is an elongated barrel-shaped drum found predominantly in South India. It is derived from another instrument called the pakhavaj and is used as the primary rhythmic accompaniment in Carnatic music as well as in religious Kirtan music. In the east (Bengal, Orissa), this barrel-shaped drum is known as the khol.

Figure 1.3.1. Music Instrument: Mridangam.

Flute

Flute is a simple cylindrical tube of uniform bore and associated with Indian music since time immemorial. Flutes vary in size. Flute is held horizontally and is inclined downwards when it is played. To produce sound or melody one has to cover the finger holes with the fingers of the left and right hand.

Variations in pitch are produced by altering the effective length of the air column. Notable flute exponents are Pt. (*pandit*) Pannalal Ghosh and Pt. Hari Prashad Chaurasia. The Murali is a transverse flute made of bamboo. It is used in a variety of musical genres and is often associated with the Hindu deity Krishna.

Figure 1.3.2. Music Instrument: Flute.

Pakhavaj

The pakhavaj is a barrel-shaped drum with two heads, each contains tuning paste, or *siyahi*. The history of the pakhavaj is unknown, yet as the predecessor of both the Hindustani tabla drums and the mridangam of Karnatak music, it served as the primary accompaniment for much of Indian classical music. It appears in the musical iconography of Hindu religious painting and in the artworks of the royal Muslim courts of the Mughal Empire.

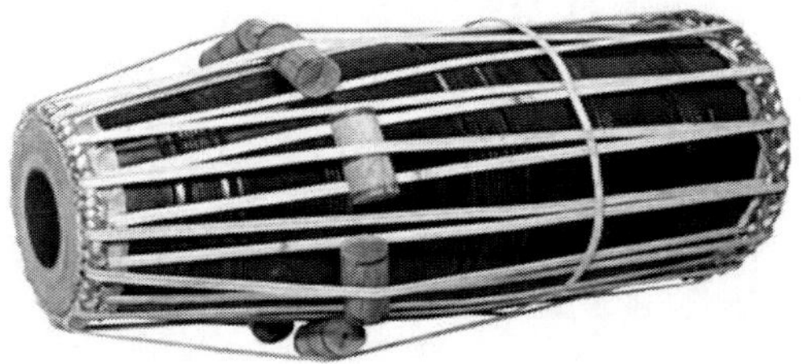

Figure 1.3.3. Music Instrument: Pakhavaj.

Rabab

The rabab is a stringed instrument with a skin-covered resonator that can be bowed or plucked depending on performance tradition. It is found in various forms throughout North Africa, the Near East, South Asia, and Central Asia. Similar to the way the Veena was adapted to eventually become what is known today as the sitar, the rabab was adapted to become the sarod. However, there are many musicians in India today who still play the rabab and it is quite popular in several music genres.

Figure 1.3.4. Music Instrument: Rabab.

Sahnai (Shenai)

The sahnai is a double reed instrument of North India and Nepal. In South India, a double reed instrument called the nagasvaram is used. Both instruments have seven equidistant fingerholes and no thumbhole. Frequently, the instrument's flared open end is made of metal while its body is made of wood or bamboo; however, they are not exclusively made in this fashion.

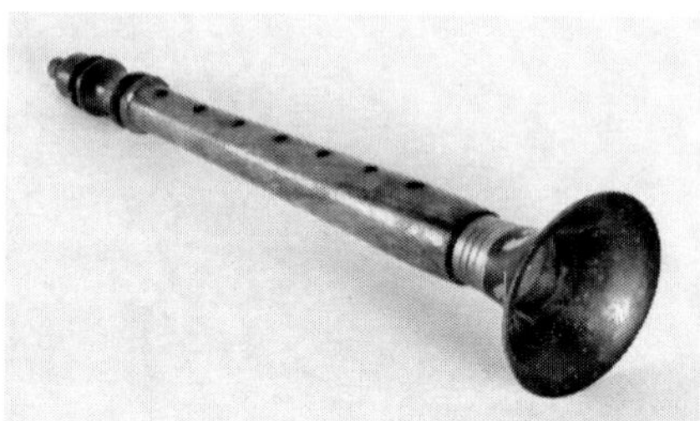

Figure 1.3.5. Music Instrument: Shenai.

Sarangi

A sarangi is a bowed stringed instrument with a skin-covered resonato. The typical sarangi is made by hand, usually from a single block of *tun* wood about 66 to 69 centimeters long. The three playing strings are made of goat gut, and the sympathetic strings (usually as many as thirty-six, though the number varies) of brass and/or steel. However, the design of sarangis varies from region to region. For example, the Nepalese sarangi is generally much smaller than its Indian counterpart, and not all sarangis have sympathetic strings.

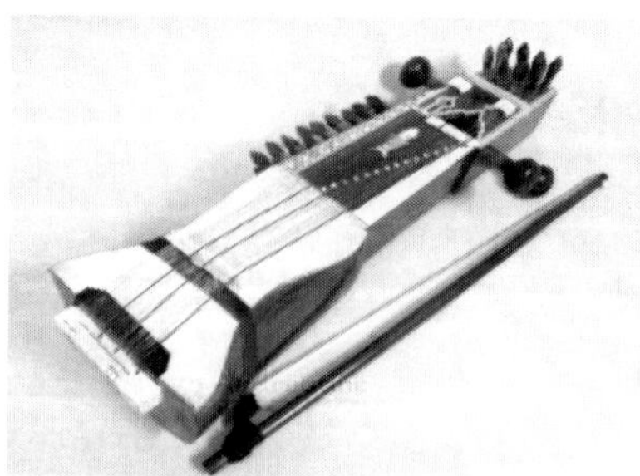

Figure 1.3.6. Music Instrument: Sarangi.

Sarod

The sarod is a relatively new instrument to South Asia, having been around for less than 200 years. The sarod is a plucked stringed instrument with a skin-covered resonator and sympathetic strings. Like the sitar, it is primarily used in Hindustani music and is accompanied by the tabla.

Figure 1.3.7. Music Instrument: Sarod.

Sitar

The sitar is easily India's most famous musical instrument overseas, having been popularized in the West by George Harrison of the Beatles, who studied with Pandit Ravi Shankar, one of the greatest sitarists of the twentieth century.

The sitar has its roots in both the Persian setar as well as in the veena. Like many stringed instruments used in classical Indian music, the modern sitar has sympathetic strings that sound only when one of the primary strings is struck on the same note. These strings, which are never played by the performer, resound in sympathy with the playing strings, creating a polyphonic timber that many have come to associate with India through the popularity of this instrument. It is interesting to note, however, that the addition of the sympathetic strings is a relatively recent development in Indian music starting in the late nineteenth century.

The use of sympathetic strings is known to have existed in other parts of the world prior to their initial use in India.

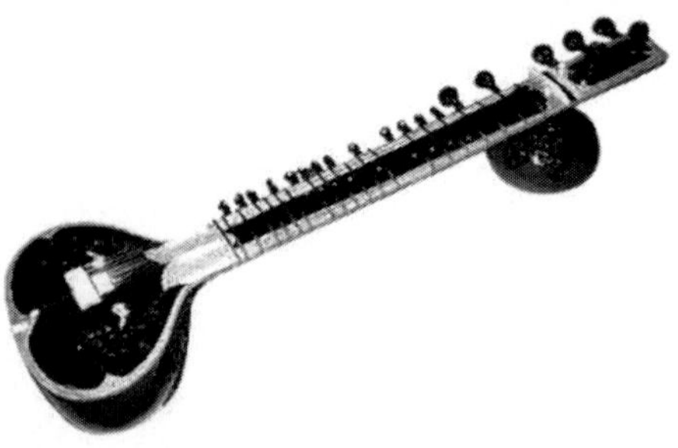

Figure 1.3.8. Music Instrument: Sitar.

Tabla

The tabla consists of two drums played by the same performer. Both drums have compound skins onto which a tuning paste, or *siyahi*, is added to help generate the wide variety of tones these drums can produce. The bayan is the larger of the two drums and is generally made of metal or pottery. The *siyahi* on the bayan is off-center, which allows the performer to add variable pressure on the skin, changing the pitch of the instrument with the palm of his or her hand while striking it with the fingertips. The smaller drum is called the dahini, or sometimes referred to as the tabla. Dahini are usually made of heavy lathe-turned rosewood and provide much higher pitch sounds than does the bayan.

Figure 1.3.9. Music Instrument: Tabla.

Tambura

The tambura is a long, stringed instrument made of light hollow wood, with either a wooden or a gourd resonator. It is typically used in accompaniment with other instruments, providing a drone pitch. India has a long history of creating musical instruments as decorative objects, and that tradition is represented in the Museum's collection.

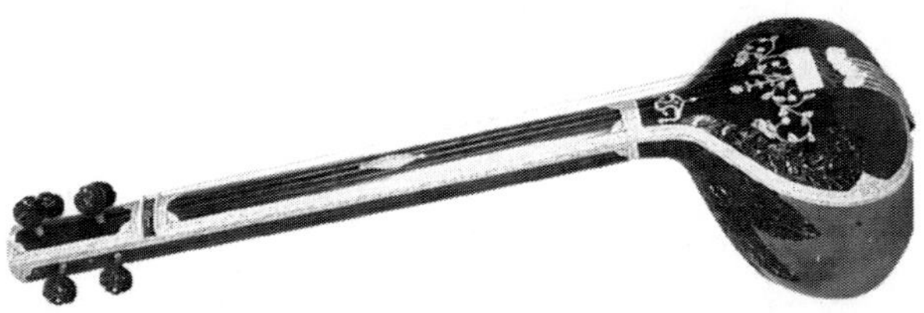

Figure 1.3.10. Music Instrument: Tambura.

Veena

Along with the pakhavaj, the Veena is one of the most commonly depicted instruments in Indian iconography. The Veena has taken many forms in both South and North India. In North India, it was called the bin or the rudraveena, and was the predecessor of the sitar. It was often built of two large gourd resonators connected by a piece of bamboo, with frets held on with wax. Most of the veenas depicted in iconography are rudraveenas. In the South, the veena—or saraswati veena—continues to be the most popular stringed instrument in classical music. In its basic shape, the Veena is a hollow wooden stringed instrument with two gourd resonators (though there can often be more than two or sometimes only one gourd resonator). The gottuvadyam, or chitraveena, is another important instrument in Carnatic music. Unlike the rudraveena and the saraswati veena, the gottuvadyam has no frets and is played with a slide using a method similar to that of the Hawaiian slide guitar.

Figure 1.3.11. Music Instrument: Veena.

Kanjira (Khanjari)

The kanjira is a frame drum of South India. It consists of a skin (usually iguana) stretched and pasted on a circular wooden frame. There are often three or four slots in the side of the frame, in which bell-metal jingle-disks are suspended from metal crossbars. The name kanjira is related to the khanjari and kanjani of North and East India and Nepal. The kanjira is tuned to various pitches by wetting the skin. It is held at the bottom of the frame by the left hand, which also varies the tension of the skin, and is beaten with the fingers of the right hand.

Figure 1.3.12. Music Instrument: Kanjira.

Kemancheh

The kemancheh is one of the world's earliest known bowed instruments. It has been altered and changed as it has traveled to other parts of the world.

Some argue that the kemancheh is the predecessor of many other stringed instruments such as the rabab, the sarangi, and the Chinese erhu.

Figure 1.3.13. Music Instrument: Kemancheh.

REFERENCES

Bharath Gupt (n.d.). *An overview of Indian Music.* Retrieved from, http://www.ifih.org/AnOverviewofIndianMusic.htm

Carnatic Music (n.d.). Retrieved from, http://en.wikipedia.org/wiki/ Carnatic_music

Carnatic Music (n.d.). Retrieved from, http://www.carnaticindia.com/ carnatic_music.html

Courtney, D. (2012). *Indian Instruments.* Retrieved from, http://chandrakantha.com/

Courtney, D. (2012). *Overview of Indian classical music.* Retrieved from, http://chandrakantha.com/articles/indian_music/

Cultural India: *Musical Instruments* (n.d). Retrieved from, http://www.culturalindia.net

Deva, B. C. (1980). *Indian Music, Indian Council for Cultural Relations,* New Delhi.

Deva, B. C. (1981). *The Music of India: A Scientific Study.* Munshiram Manoharlal Publishers Pvt. Ltd., New Delhi.

Gautam, M. R. (1980). *The Musical Heritage of India.* Delhi: Abhinav Publications.

Grout, D. J., Burkholder, J. P. & Palisca, C. V. (2010). *A History of Western Music* (8th Edn). New York: W. W. Norton.

Indian Classical Music (n.d.). Retrieved From, http://en.wikipedia.org/wiki/Indian_classical_music

Musical Instruments of the Indian subcontinent (n.d.). Retrieved from, http://www.metmuseum.org/

Scaruffi, P. (2002). *A brief summary of Indian music.* Retrieved from, http://www.scaruffi.com/history/indian.html

Ranade, G. H. (1971). *Hindusthani Music- Its Physics and Aesthetics.* Popular Prakashan, Bombay

Seaton, D. (2010). *Ideas and Styles in the Western Musical Tradition* (3rd edn). United Kingdom: Oxford University Press.

Swami prajnananda, (1965).*A Historical Study of Indian Music.* Calcutta. Anandadhara Publication.

Western music (n.d.). Retrieved from, http://en.wikipedia.org/wiki/Western_music

Western Music (n.d.). Retrieved from, http://www.britannica.com/EBchecked/topic/398976/Western-music

Western music history (n.d.). Retrieved from, http://en.wikibooks.org/wiki/Western_Music_History

Chapter 2

PARAMETERS OF VOICE PRODUCTION

The human voice is a tool for communication and a vocal instrument at the same time. Its study is complex. The voice can be characterized by physiological, acoustic and psychological aspects. For a vocalist, voice is the most important investment that earns him/her popularity and trade. It is imperative for these elite performers to conserve, prevent abuse and maintain vocal health to meet their occupational requirements. In India, scientific examination of classical singing and other types of singing (light music, folk music, music associated with dance, etc.) is still a relatively new concept, despite the traditional styles and pedagogies, high cultural relevance and vocal health issues. It is important to understand acoustics and physiology of voice production to help develop training methods that facilitate the mastery of skilled singing.

Culturing the voice depends on the inherent texture of the voice and the genre of singing as in classical (Indian/ Western), folk and light. The singing voice and the voice with which we speak aren't the same. While speaking is an action that doesn't need any conscious effort (do not refer to speaking as in stage orations), singing does demand the same. The technique of bringing the voice under our command is termed as "voice culture". This comes with breath control techniques, healthy body and more importantly a sound and focused mind.

While for an instrument, oiling, tuning, selecting the instruments made out of the right materials (depending on the instrument kind - percussion, string, air instrument) etc. would matter, culturing a voice needs an ardent effort and a personal care. Every voice is unique and beautiful. The voice needs to be understood by its owner to nourish and culture it in a way that it gets strengthened. Each voice has its traits and limitations. Voice culture must aim at reducing the shortcomings and enhance the impressive traits of the voice.

THE HUMAN VOICE

Of the innumerable instruments, very few can match the most natural instrument - the voice. Carnatic music gives pride of place to Vocal music. This is because vocal music has the added dimension of lyrics, which is one of the basic components of Carnatic music. A vocalist can project the lyrics and the theme of the music the best. Even the melody instruments try to approximate to vocal standards. Hence, a student interested in instrumental career generally learns vocal first and then repeats the instrumental music.

Voice Mechanisms

Singing can be defined as the musical expression of emotion through the medium of vocal organs and the organs of speech. The technique of voice production for singing is more complex than it is for speech, as this requires the control of three sets of muscles - inspiration and expiration (respiratory muscles), phonation (intra and extra-laryngeal muscles) and those of articulation (muscles of tongue, jaw, lips and soft palate).

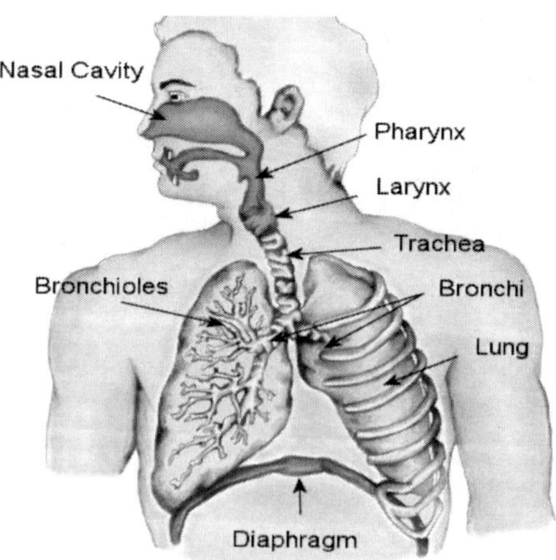

Figure 2.1. Structures involved in respiration.

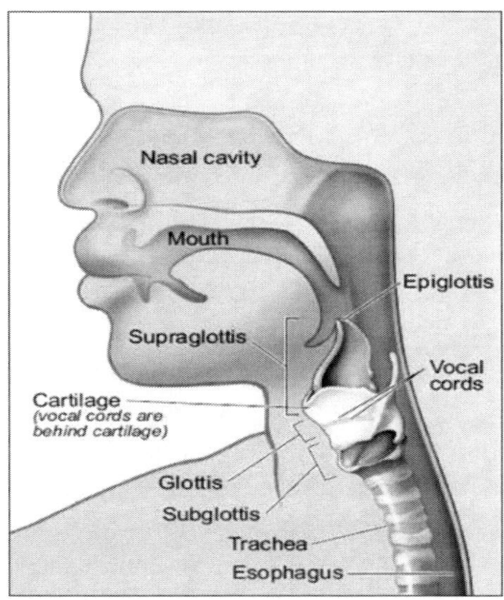

Figure 2.2.Structures of articulatory and phonatory systems.

Vocal Techniques

A Carnatic vocalist is expected to possess a voice that is rich in tone and volume, has depth and which is capable of sustaining different notes for long periods without any wobble. He/she must also possess a range of at least two and a half octaves and execute with clarity and verve, phrases of different tempo. The various embellishments or ornamentations (gamakas) and tonal shades should be aptly produced for rendering different types of musical compositions and other creative aspects of Carnatic music. The technical exercises and compositions of Carnatic music are designed to impart all the above. Of course, the student must have the right attitude, technical guidance and perseverance!

Singing when done with proper vocal technique is an integrated and coordinated act that effectively coordinates the physical processes of singing. There are four physical processes involved in producing vocal sound: respiration, phonation, resonation, and articulation. These processes occur in the following sequence:

1.   Breath is taken into the body.
2.   Sound is initiated in the larynx

3. The vocal resonators receive the sound and influence it
4. The articulators shape the sound into recognizable units

Although these four processes are often considered separately when studied, in actual practice they merge into one coordinated function. With an effective singer or speaker, one should rarely be reminded of the process involved as their mind and body are so coordinated that one only perceives the resulting unified function. Many vocal problems result from a lack of coordination within this process.

Since singing is a coordinated act, it is difficult to discuss any of the individual technical areas and processes without relating them to the others. For example, phonation only comes into perspective when it is connected with respiration; the articulators affect resonance; the resonators affect the vocal folds; the vocal folds affect breath control; and so forth. However, some areas of the art of singing are so much the result of coordinated functions that it is hard to discuss them under a traditional heading like phonation, resonation, articulation, or respiration.

Once the voice student has become aware of the physical processes that make up the act of singing and of how those processes function, the student begins the task of trying to coordinate them. Inevitably, students and teachers will become more concerned with one area of the technique than another. The various processes may progress at different rates, with a resulting imbalance or lack of coordination. The areas of vocal technique which seem to depend most strongly on the student's ability to coordinate various functions are:

1. Extending the vocal range to its maximum potential
2. Developing consistent vocal production with a consistent tone quality
3. Developing flexibility and agility
4. Achieving a balanced vibrato

Developing the Singing Voice

Singing is a skill that requires highly developed muscle reflexes. Singing does not require much muscle strength but it does require a high degree of muscle coordination. Individuals can develop their voices further through the careful and systematic practice of both songs and vocal exercises. Vocal pedagogists instruct their students to exercise their voices in an intelligent manner. Singers should be thinking constantly about the kind of sound they are making and the kind of sensations they are feeling while they are singing. Vocal exercises have several purposes, including warming up the voice; extending the vocal range; "lining up" the voice horizontally and vertically; and acquiring vocal techniques such as legato, staccato, control of dynamics, rapid figurations, learning to sing wide intervals comfortably, singing trills, singing melismas and correcting vocal faults.

Extending Vocal Range

An important goal of vocal development is to learn to sing to the natural limits of one's vocal range without any obvious or distracting changes of quality or technique. Vocal pedagogists teach that a singer can only achieve this goal when all of the physical processes involved in singing (such as laryngeal action, breath support, resonance adjustment, and articulatory movement) are effectively working together. Most vocal pedagogists believe in coordinating these processes by (1) establishing good vocal habits in the most comfortable tessitura of the voice, and then (2) slowly expanding the range.

There are three factors that significantly affect the ability to sing higher or lower:

1. The *energy* factor – "energy" has several connotations. It refers to the total response of the body to the making of sound; to a dynamic relationship between the inhalatory muscles and the exhalatory muscles known as the breath support mechanism; to the amount of breath pressure delivered to the vocal folds and their resistance to that pressure; and to the dynamic level of the sound.
2. The *space* factor – "space" refers to the size of the inside of the mouth and the position of the palate and larynx. Generally speaking, a singer's mouth should be opened wider the higher he or she sings. The internal space or position of the soft palate and larynx can be widened by relaxing the throat. Vocal pedagogists describe this as feeling like the "beginning of a yawn".
3. The *depth* factor – "depth" has two connotations. It refers to the actual physical sensations of depth in the body and vocal mechanism, and to mental concepts of depth that are related to tone quality.

McKinney states that the above mentioned factors can be expressed in three basic rules: (1) As one sings higher, he/ she must use more energy; as one sings lower, he/she must use less. (2) As one sings higher, he/she must use more space; as one sings lower, he/she must use less. (3) As one sings higher, he/she must use more depth; as one sings lower, he/she must use less.

Posture

The singing process functions best when certain physical conditions of the body are put in place. The ability to move air in and out of the body freely and to obtain the needed quantity of air can be seriously affected by the posture of the various parts of the breathing mechanism. A sunken chest position will limit the capacity of the lungs, and a tense abdominal wall will inhibit the downward travel of the diaphragm. Good posture allows the breathing mechanism to fulfill its basic function efficiently without any undue expenditure of energy. Good posture also makes it easier to initiate phonation and to tune the resonators as proper alignment prevents unnecessary tension in the body. Vocal pedagogists have also noted that when singers assume good posture it often provides them with a greater sense of self-assurance and poise while performing. Audiences also tend to respond better to singers with good posture. Habitual good posture also ultimately improves the overall health of the body by enabling better blood circulation and preventing fatigue and stress on the body.

There are eight components of the ideal singing posture:

1. Feet slightly apart
2. Legs straight but knees slightly bent
3. Hips facing straight forward
4. Spine aligned
5. Abdomen flat
6. Chest comfortably forward
7. Shoulders down and back
8. Head facing straight forward

Breathing and Breath Support

Natural breathing has three stages: breathing-in period, a breathing out period, and a resting or recovery period; these stages are not usually consciously controlled. Within singing there are four stages of breathing: a breathing-in period (inhalation); a setting up controls period (suspension); a controlled exhalation period (phonation); and a recovery period.

These stages must be under conscious control by the singer until they become conditioned reflexes. Many singers abandon conscious controls before their reflexes are fully conditioned which ultimately leads to chronic vocal problems.

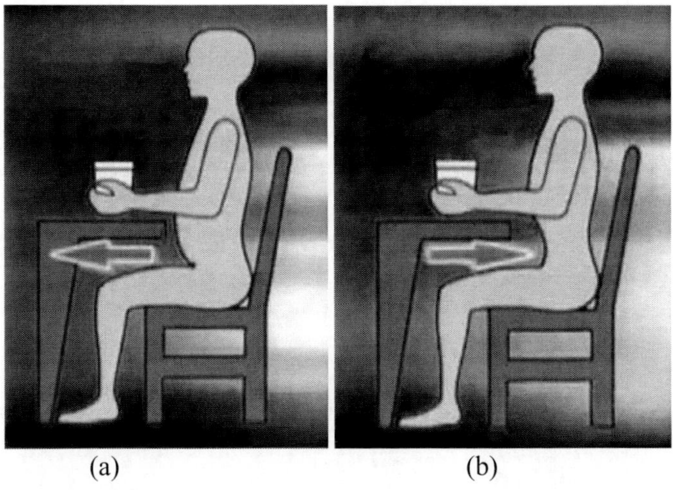

(a)   (b)

Figure 2.3. (a) Inhalation in abdominal breathing shows outward movement of the abdomen as air is filled in the lungs; (b) Exhalation in abdominal breathing shows inward movement of the abdomen as air is expelled from the lungs.

Breathing patterns have been basically classified into categories, as in, *Clavicular/ high breathing,* wherein maximum amount of effort is required to obtain a minimum amount of benefit; *Intercostals / Mid breathing,* where mainly the middle parts of lungs are filled with air, and the lung capacity is not utilized fully, and; *Diaphragmatic or low breathing,* wherein more air is taken while inhaling, due to greater movement of lungs. Diaphragmatic breathing, also referred to as abdominal breathing, utilizes maximum capacity of the lungs, and thus is most suitable for activities such as singing and other physical activities as running.

Vibrato

Vibrato is used by singers, in which a sustained note wavers very quickly and consistently between a higher and a lower pitch, giving the note a slight quaver. Vibrato is the pulse or wave in a sustained tone. Vibrato occurs naturally, and is the result of proper breath support and a relaxed vocal apparatus. Some singers use vibrato as a means of expression. Many successful artists have built a career on deep, rich vibrato.

VOICE CULTURE

Good music is created when "Voice" produces the sound at the correct pitch, accurate loudness and with good quality. Voice culture is the science of perfecting and using one's voice effectively. It is also called science of the human voice. "Voice" is also an instrument for a vocalist to communicate his/her musical ideas. "It has to be kept finely tuned. And Voice –Training and Voice –Culture helps to improve its quality, its power and endurance. Voice is the only instrument that can be cultivated, improved and cultured, by variation of pitch, intensity and timbre (tonal quality). In Sangeeta Ratnakar, *"sareera"* is a term which refers to a gifted voice without any regular practice. Sareera is a "natural voice". The Voice –training method turns the gifted natural instrument into an exquisite and melodious "Vadhyam" or instrument.

Voice is an only instrument that can be cultivated, improved and cultured by variation of pitch, intensity and timbre (tonal quality). Indian Classical Music is based on spirituality. For a vocalist of Indian Classical Music "Voice" is to sing the glory of God. Any determined effort focused towards attaining perfection is called 'Sadhana'. A vocalist has to exercise and make the vocally generated music worthy towards attaining perfection and is called 'Kantha–Sadhana'. In Western Music it is called "Voice-Culture". Indian Classical Music is always connected with spirituality. When a vocalist of Indian Classical Music sings in correct tone, the whole environment becomes spiritual and soothing. It seems to pour forth freely from the Heart spontaneously. There is no evidence of calculation, of carefully directed effort, of attention to the workings of the voice in the tones of a perfect singer.

In Indian Classical music(both Hindustani and Carnatic), the teaching of Voice production has been purely on empirical basis so far, relying on the pernicious system of trial and error, with the result that this branch has been exposed to more charlatanism.

A proper understanding of the structure and foundation of the physiological mechanism, acoustical principles of voice emission and psychological factors of the individual help to deduce scientific facts which would guide through us the tangle of confusion which has beset the art of Voice Culture. Although the classical system of music has been enriched with many exercises to develop the skill of singing, a comprehensive study of all the scientific aspects related to voice cultivation and also the practice of technically developed exercises will help the singers to analyze their own voice and to manipulate the vocal apparatus for the perfect voice modulation.

There is a very important fact that an effective and impressive voice is a must for singers to succeed. That is why Voice Culture is considered very important for a singer. One interesting thing is that it is helpful for other vocally-active professionals like public speakers, actors, and teachers. Conceptually, an ideal voice differs according to the System of Music. Western music is based on Harmony while Indian music is based on Melody. Carnatic music has its own unique style whereas in Hindustani music every Gharana's voice training is different. Soft singing is necessary in film music. But the best voice is one which is most flexible.

Voice training has become imperative in the current scenario; it helps Voice flexibility and ensures Control of Breath, Uniformity in voice, Clarity in voice, Articulation, Resonance and Vocal range. The scientific and medical aspect of voice culture is to maintain Voice hygiene and preserve the Voice from Voice disorder. As is the case with most other systems of the world, Carnatic music also uses the human voice as well as several instruments of both Indian and Western origin. There is an amazing open-mindedness when it comes to adopting good things from other systems. Several instruments like the violin, mandolin, guitar and the saxophone have been adopted successfully, of course, with a few modifications to suit the requirements of Carnatic music.

DO's and DON'Ts

- Practicing early in the morning, soon after waking up may not suit everyone. Give sufficient time to your voice to even out. Warm water gargling would help. Talk to people. Get your tongue to movement and leave your voice chords moist before you start the first practice for the day.

- DO NOT practice at very high pitch (Shruthi) when the sun is right on head (that's between 12 and 4) i.e., in the noon. If your pitch is five and half, it will do good to stick to five (say half a pitch lower than yours) at this time of the day. This would avoid wear and tear.

- Every practice should start with a varnam preferably; varnams are structured in a way that would exercise the voice from all possible angles.

- Learn to modulate the voice. DO NOT sing with the maximum voice beyond upper *sthayi rishabham*. Soften the voice to reduce the stress on the vocal cord. This doesn't mean you bring in false voice. Using false voice may help in effortless singing but wouldn't reach the audience in effect. Know the difference between singing with open voice and singing loudly and the difference between singing with a false voice and singing softly.

- Proper physical posture is required for a good voice production. Certain voice types would get strained with 2-3 hours of practice but some people can go up to 5-6 hours in one stretch. It is important that you understand the nature of your voice and its limitations. One should know his/her own competence to practice and NOT to overdo! If the posture and the singing technique followed are acceptable, at the end of the practice session, the abdominal muscles should get strained and not the voice chords. This could be a self check. (Nabhi hruth khanta rasana- Origin of voice is from the abdomen).

- Nasal touch is required for a proper alignment with Shruthi, but nasal twang shouldn't inhabit your voice completely. Open mouth singing helps reduce the nasal twang.

- Diet plays an important role in maintaining the voice health. Drinking warm water helps; hot tea, clove in honey, ginger kashayam (tonic), and pepper milk do wonders and soften the voice. Avoid road-side food, oily food, and cold items.

- Avoid overuse of the voice- don't scream, talk softly and talk in moderation. Last and the most importantly, get rid of the fear of falling sick! Have a strong mind and a committed soul.

Figure 2.4.Representations of a correct posture in the sitting position.

REFERENCES

Aronson, A. E., & Bless, D. (2009). *Clinical Voice Disorders* (4th ed.). New York: Thieme Medical.

Boominathan, P., Rajendran, A., Nagarajan, R., Jayashree, S., & Muthukumaran, G. (2008). Vocal abuse and vocal hygiene practices among different level professional voice users in India- A survey. *Asia Pacific Journal of Speech Language and Hearing, 11,1,* 47-53.

Chapman, J. L. (2006). *Singing and teaching singing: A holistic approach to classical voice.* San Diego, Calif: Plural Publishing.

Deutsch, D. (2013). *The Psychology of Music* (3rd ed.). San Diego: Elsevier.

Davies, G. & Jahn, A. F. (1998).*Care of the professional voice.* Boston:Oxford.

Ravikumar, A., Boominathan, P., & Mahalingam, S. (2010). Clinical voice analysis of Carnatic singers. *Proceedings of the 4rth World Voice Congress, Seoul, Korea,* pp. 173.

Sataloff, R. T. (1997). *Professional Voice: The science and art of clinical care* (2nd ed.). San Diego: Singular Publishing.

Zemlin, W. R. (2010). Speech and Hearing Science: Anatomy and physiology (4[th]ed.). Boston: Allyn and Bacon.

Chapter 3

THE HUMAN EAR

Anatomy and Physiology

The auditory system can be majorly divided into the peripheral auditory system and the central auditory nervous system. Peripheral system (Figure 3.1) includes; outer, middle and inner ear while the auditory nervous system comprises of auditory nerve and various relay centers starting from cochlear nucleus to auditory cortex.

Peripheral Auditory System

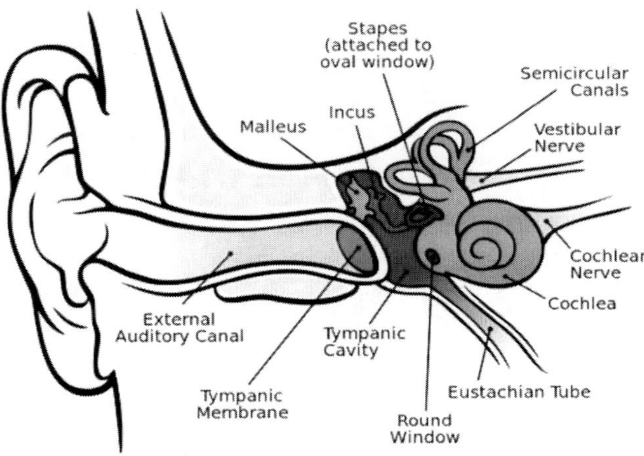

Figure 3.1.Major parts of the peripheral auditory system.

Outer Ear

- ***Pinna***

The pinna also known as auricle (Figure 3.2) is the external appendage of the ear. It is an irregularly shaped ovoid of highly variable size, which is basically composed of skin-covered elastic cartilage. It contains some grossly undifferentiated extrinsic muscles that are of a completely vestigial nature in humans. Outer part of pinna demarcated by a ridge-like rim is called the helix. If we first follow the helix posteriorly from the top of the ear to end we get earlobe (lobule) at the bottom. Unlike the rest of the pinna, the lobe does not have any cartilage.

Starting again from the apex of the helix, we see that it proceeds anteriorly and downward, and then turns posteriorly in a rather sharp angle to form the crus of the helix. The scaphoid fossa is a depression lying between the posterior portion of the helix posteriorly and a ridge called the antihelix anteriorly. The antihelix is a ridge that runs essentially parallel to the helix. Its upper end bifurcates to form two crura, rather wide superoposterior crus and narrower anterior crus, which ends under the angle where the helix curves backward.

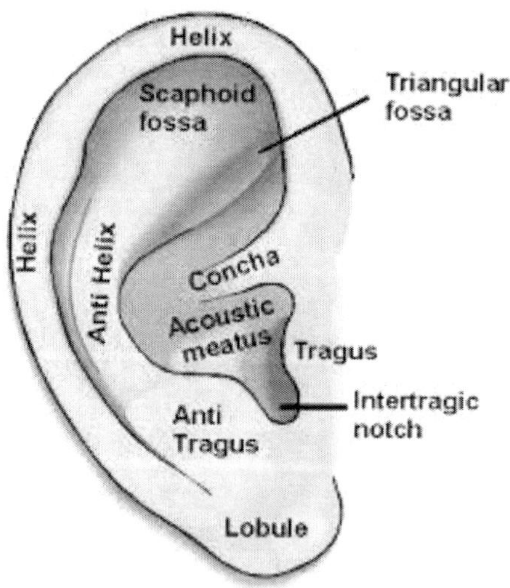

Figure 3.2.Important landmarks of the Pinna.

A triangular depression is thus formed by the two crura of the antihelix and the anterior part of the helix, and is called the triangular fossa. From the crura, the antihelix curves downward and then forward, and ends in a mound-like widening, the antitragus. Opposite to the antitragus is a backward-folding ridge called the tragus. The acute angle formed by the tragus and antitragus is called the intertragal notch. The tragus, the antitragus, and the crus of the helix border a relatively large and cup-shaped depression called the concha which marks entry to canal.

From the auditory point of view, the pinna is mainly involved in monaural localization of sounds. The ridges and depressions in pinna are believed to be contributing to medial plane localization of sounds. The concha has also been reported to function as cavity resonator and its resonance property has been found to produce a gain of approximately 10dB at 4 to 5 kHz and a loss of approximately 5dB at 10 kHz. A non-auditory function of pinna is probably to protect ear canal from foreign bodies as tragus covers ear canal opening.

- ***Ear Canal***

The ear canal also called as external auditory meatus (EAM), leads from the concha to the eardrum and varies in both size and shape. The outer portion of the canal, about one-third of its length, is cartilaginous; the remaining two-third is bony. The EAM has a length of approximately 2.5cm with a diameter of approximately 0.6 cm and is somewhat S-shaped. It is for this reason that the pinna must be pulled up and back in order to visualize the eardrum.

The protective function of the EAM is directly related to its physical shape and properties. The S-shape course, depth and the rigidity of the walls, protect ear drum from direct injury. Outward facing hairs in the cartilaginous part of the canal hinder the entry of invaders and thus protect ear drum indirectly. The EAM is closed at one end by ear drum and open at the other end. In such a tube, according to quarter wave theory, resonation occurs when the length of the tube is one fourth the wave length of the applied sound. The average length of the ear canal is 2.3cm and thus approximately at 3.6 kHz resonance effect can be observed. In children as the length of the tube is shorter, resonant frequency will be higher. Studies have shown that gain given to high frequencies by ear canal through this resonance effect is approximately 15dB.

Middle Ear

The middle ear cavity or tympanum (Figure 3.3) can be imagined to be a six-sided room. It consists of the ear drum that terminates the ear canal and the three small bones (ossicles); the malleus, the incus and the stapes. Two small muscles, the tensor tympani muscle and the stapedius muscle, are also located in the middle ear. The Eustachian tube/Auditory Tube connects the middle ear cavity to the pharynx.

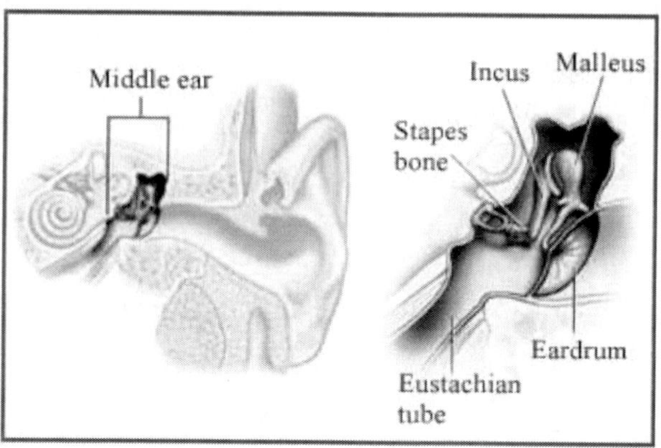

Figure 3.3.Middle ear cavity showing ossicles.

Tympanic Membrane

The canal terminates at the eardrum also called as tympanic membrane (Figure 3.4) which is a slightly oval, thin membrane. It is tilted laterally at the top, so as to sit in its annulus at an angle of about 55 degree to the ear canal. If seen from concha, it is concave in shape. The tympanic membrane is divided into the pars flaccid above and the pars tensa below. One of middle ear bones, the malleus is attached to tympanic membrane at a position called umbo.

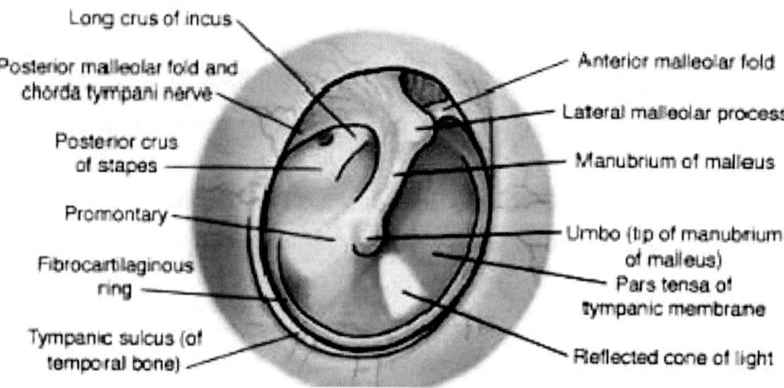

Figure 3.4.Tympanic membrane showing major parts.

- ***Ossicles***

There are three ossicles in each ear; malleus, incus, and stapes; they are connected to each other and collectively referred to as the ossicular chain. The malleus is commonly called the hammer. It is the largest of the ossicles, being about 8 to 9mm long and weighing approximately 25mg. The incus weighs approximately 30 mg and has a length of about 5 mm. The stapes (stirrup) is the smallest of the ossicles. It is about 3.5 mm high, and the footplate is about 3 mm long by 1.4 mm wide. It weighs on the order of 3 to 4 mg.

The eardrum is merged to the malleus, which connects to the incus, which in turn connects to the stapes. Vibrations of the stapes footplate introduce pressure waves in the inner ear. The movement of the ossicles may be stiffened by two muscles. The stapedius muscle, the smallest skeletal muscle in the body, connects to the stapes and is controlled by the facial nerve; the tensor tympani muscle connects to the base of the malleus and is under the control of the medial pterygoid nerve which is a branch of the mandibular nerve of the trigeminal nerve. These muscles contract in response to loud sounds, thereby reducing the transmission of sound to the inner ear.

A principle in acoustics states that sound waves travelling in a medium of some given density and elasticity will not pass readily into a medium with different elasticity and density but, rather most of the sound will be reflected away. When environmental sound travelling through the air enters the outer ear and directly reaches inner ear which contains fluids, most of the energy gets reflected back due to impedance mismatch. Middle ear plays an important role here by avoiding such a transmission loss and instead acting as a mechanical transformer that boosts the original signal so that the energy can be efficiently transmitted to the cochlea. Combination of three mechanisms related to middle ear viz. the area ratio advantage of ear drum to the oval window, the curved membrane buckling effect of the tympanic membrane and the lever action of the ossicular chain act together in providing a total advantage of about 38dB.

- ***Eustachian Tube***

The Eustachian tube is also known as the auditory tube. The overall length of the tube is about 3.5 cm. The first one third of the Eustachian tube beginning at the middle ear is surrounded by bone, whereas the remainder is enclosed within an incomplete ring of hook-shaped elastic cartilage, ending near throat. Four muscles involved in supporting the function of Eustachian tube include, Tensor Velipalatani, Levator Velipalatani, Salphingopharyngous & Tensor tympani. Three main functions of this tube include, pressure equalization between middle ear and atmosphere, protection from potential deleterious effects of loud sounds related to non-acoustic functions like phonation and drainage of secretions from middle ear to the nasopharynx through an active mucociliary system.

Inner Ear

The inner ear is not a chamber like the middle ear, but consists of several tubes which wind in various paths within the skull. The inner ear structures are contained within bony labyrinth (Figure 3.5A) which is grossly divided into three sections: the vestibule, the cochlea, and the semicircular canals. The membranous labyrinth (Figure 3.5B) which contains the end organs of hearing and balance is enclosed within the bony labyrinth.

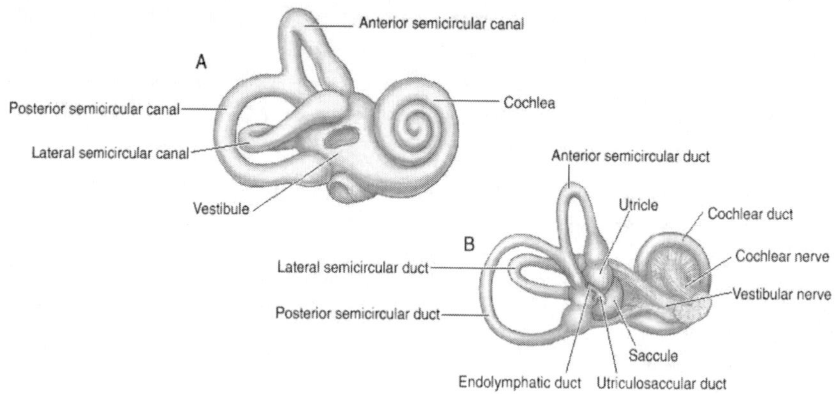

Figure 3.5.Bony labyrinth (A) and Membranous labyrinth (B).

The cochlea, which is the end organ for hearing function, is a snail-shaped structure having a little more than 2 ½ turns. When uncoiled, the cochlea has a length of 3.1–3.3 cm. It has three fluid-filled canals (Figure 3.6): the scala vestibule (Vestibular canal); the scala tympani (Tympanic canal); and the scala media (Cochlear duct). The scala media, located in the middle of the cochlea, is separated from the scala vestibuli by Reissner's membrane and from the scala tympani by the basilar membrane. The Basilar membrane is widest (0.42–0.65 mm) and least stiff at the apex of the cochlea, and narrowest (0.08–0.16 mm) and most stiff at the base. High-frequency sounds localize near the base of the cochlea while low-frequency sounds localize near the apex. The organ of Corti (Figure 3.6), located along the basilar membrane, contains the sensory cells (hair cells) that transform the vibration of the basilar membrane into a neural code. There are roughly 12,000 outer hair cells (OHC) and 3500 inner hair cells (IHC) in each ear.

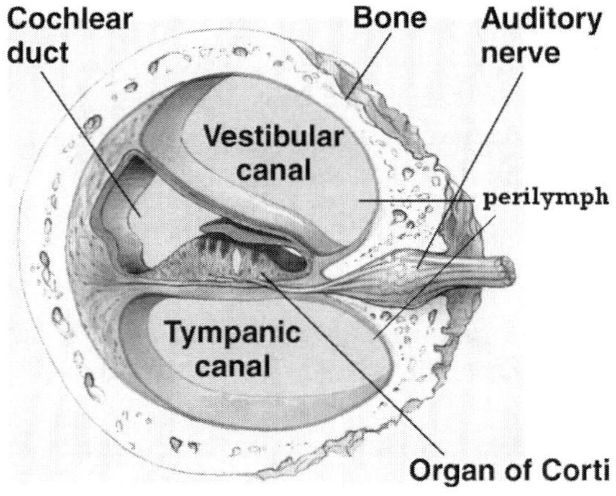

Figure 3.6.Cross section of the cochlear turn showing three divisions and organ of corti.

Stimulation is transmitted to the cochlear fluids by the in-and-out motions of the stapes foot plate at the base of the cochlea. This movement displaces fluid in the inner ear and thus moving basilar membrane upward and downward in accordance to phase of the sound pressure. Sound transmitted to the cochlea develops a special pattern of wave called travelling wave and the movement is best explained by travelling wave theory. Different region of the basilar membrane gets activated by different frequency sounds and this is explained on the basis of gradation of stiffness on the basilar membrane (apex being least still while base most stiff). The travelling wave moves the sensory hair cells on specific location on the basilar membrane, thus activating the sensory process wherein electrical signals are generated.

AUDITORY NERVOUS SYSTEM

Auditory nervous system consists of vestibulo-cochlear nerve, also called as 8[th]cranial nerve. One part of this cranial nerve consists of the cochlear nerve. It has two types of fibers known as type I and type II. The type I nerve fibers are thought to carry all the auditory information in the form of electrical impulses from the organ of corti to higher centers of the central nervous system. The type II fibers form the outer spiral fibers which innervate outer hair cells from higher centers. The other portion of the vestibulo-cochlear nerve is the vestibular nerve, which carries balancing information to the brain from the semicircular canals.

Higher centers of the auditory system mainly include cochlear nucleus, superior olivery complex, lateral lemniscus, medial geniculate body, inferior colliculus and auditory cortex. The auditory nerve terminates in the cochlear nucleus (CN), which is the first relay nucleus of the ascending auditory pathways. It is located in the lower brainstem, at the junction between the medulla and the pons on the same side as the ear from which it receives its innervations. The two sides' cochlear nuclei are connected. The superior olivary complex (SOC) receives input from both sides' CN. The SOC is thus the first group of nuclei that integrate information from both ears.

The lateral lemniscus (LL) is the pathway from the lower nuclei of the auditory pathway. Each LL includes neural fibers originating from the CN and SOC on both sides, as well as fibers arising from the nuclei of the LL itself. The LL principally includes a ventral nucleus (VNLL) and a dorsal nucleus (DNLL); The DNLL receives input from both ears and is involved in binaural hearing while the VNLL mainly receives input from the contralateral ear. Some of the axons that lead away from the DNLL travel to the opposite side as the commissure of Probst.

The majority of the ascending fibers from the LL project to the inferior colliculi (IC), which are large nuclei on the right and left sides of the midbrain. The IC on one side connects to the IC on the other side and these connections are important for directional hearing that is based on the differences in the sound intensity at the two ears. The medial geniculate body (MGB) is the thalamic auditory relay nucleus where all fibers that originate in the IC are interrupted.

The primary auditory cortex is the part of the cerebral cortex that lies in the superior temporal gyrus of the temporal lobe and extends into the lateral sulcus and the transverse temporal gyri (also called *Heschl's gyri*). There are additional areas of the human cerebral cortex that are involved in processing sound, in the frontal and parietal lobes. The auditory cortex is composed of fields which differ from each other in both structure and function.

The primary auditory cortex processes basic and higher functions related to hearing. As with other primary sensory cortical areas, auditory sensations reach perception only if received and processed by a cortical area.

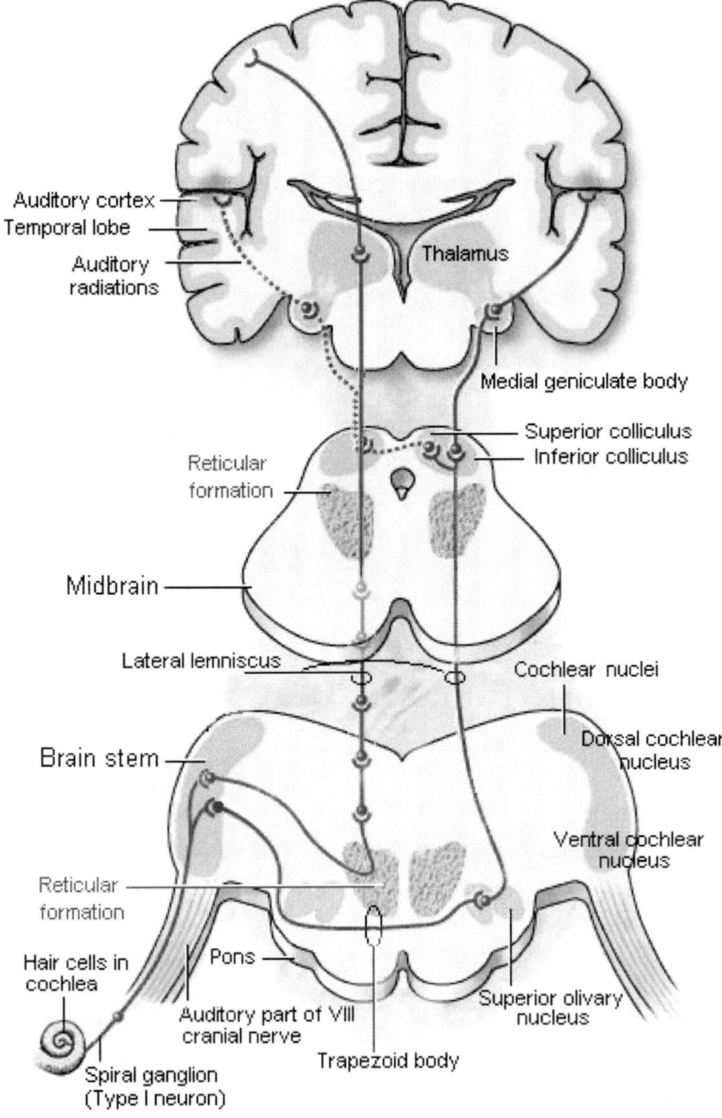

Figure 3.7. Auditory Pathway showing important relay centers.

Neurons in the auditory cortex are organized according to the frequency of sound to which they respond best. Neurons at one end of the auditory cortex respond best to low frequencies; neurons at the other respond best to high frequencies. There are multiple auditory areas which can be distinguished anatomically and on the basis that they contain a complete "frequency map."

The auditory cortex is involved in tasks such as identifying and segregating auditory "objects" and identifying the location of a sound in space.

Human brain scans have indicated that a peripheral bit of this brain region is active when trying to identify musical pitch. When each instrument of a symphony orchestra or the jazz band plays the same note, the quality of each sound is different but the musician perceives each note as having the same pitch. The neurons of the auditory cortex of the brain are able to respond to pitch.

In a nutshell:

> We are able to hear when our ears convert the vibrations of a sound wave in the air into signals that our brains interpret as sound. Our ears are divided into three parts, the external ear, the middle ear and the inner ear. Sound waves, traveling at a speed of approximately 740 miles per hour, enter the ears and are funneled through the ear opening, down the ear canal to the eardrum (or tympanic membrane). The sound waves strike the eardrum, causing it to vibrate. These movements of the tympanic membrane are transmitted across the three middle ear bones (hammer or *malleus*, anvil or *incus* and the stirrup or *stapes*), which act as a transformer, changing sound vibrations in the air into fluid waves in the inner ear. The fluid waves bend delicate nerve endings (hair cells) in the cochlea, creating electrical impulses. These electrical impulses are then transmitted by the hearing nerve (cochlear nerve) to the brain, where they are interpreted as understandable sounds.

Basic Psychoacoustical Aspects and Auditory Processing

The Range of Hearing

The intensity range a normal hearing person hears is the limit between just audible threshold and intolerable level. This range is usually described as dynamic range and it is normally around 120dB. Audio metrically, normal hearing threshold i.e. the lowest level at which a normal hearing individual hears sound at least 50% of the times is between -10 to 15dBHL. If the hearing level is above 15dB then that person is said to be having hearing loss. The frequency range for which human ears are sensitive is between 20 Hz to 20 kHz. However, hearing thresholds are reasonably sensitive between approximately 100 to 10 kHz and most sensitive (lower threshold) between 2 kHz to 5 kHz.

Differential sensitivity

The smallest perceivable difference between two sounds is referred to as just noticeable difference (jnd). For instance the smallest intensity difference that can be perceived is the jnd for intensity. At hearing threshold level jnd for loudness for normal hearing adults is about 3dB. However, it decreases with increase in intensity. In similar lines, there exists jnd for frequency and duration aspects of sound.

Loudness and Pitch

Sound being a physical stimulus has parameters like frequency and intensity. Corresponding psychological responses to such parameters are described as pitch and loudness respectively. High intensity sounds are perceived as louder. Also higher frequency sounds are perceived as having higher pitch. However, it has to be noted that there is no one to one correspondence between the physical stimuli and how they are perceived.

Localization

We are able to perceive the direction of a sound source with some accuracy. Left vs. right location is determined by perception of the difference of arrival time or difference in phase of sounds at each ear. If there are more than two arrivals, as in a reverberant environment, we choose the direction of the first sound to arrive, even if later ones are louder. Localization is most accurate with high frequency sounds with sharp attacks.

Information about height of the sound source is provided by the shape of our ears. If a sound of fairly high frequency arrives from the front, a small amount of energy is reflected from the back edge of the ear lobe. This reflection is out of phase for one specific frequency, so a notch is produced in the spectrum. The elongated shape of the lobe causes the notch frequency to vary with the vertical angle of incidence, and we can interpret that effect as height. Height detection is not good for sounds originating to the side or back, or lacking high frequency content.

Masking

The threshold of hearing for a particular tone can be raised by the presence of another noise or another tone. White noise (wide band noise comprising of almost entire audible frequency range) reduces the loudness of all tones, regardless of absolute level. If the bandwidth of the masking noise is reduced, the effect of masking loud tones is reduced, but the threshold of hearing for those tones remains high. If the masking sound is narrow band noise or a tone, masking depends on the frequency relationship of the masked and masking tones. At low loudness levels, a band of noise will mask tones of higher frequency than the noise better than those of lower frequency. At high levels, a band of noise will also mask tones of lower frequency than itself.

Auditory Processing

Auditory Processing is a natural process of taking in sound through the ear and having it travel to the language area of the brain to be interpreted. Auditory processing skills develop at different rates in children.

Auditory processing capacity is the ability of an individual to hold sequence and process or understand what he/she has heard. Hearing acuity is only one part in the processing that occurs in the auditory system. There are many people, both children and adults, who have no trouble with actual hearing acuity that is to say that the hearing is normal, but who have other types of auditory difficulties. Auditory processing is the ability to interpret the sounds that one has heard. In central auditory processing, the brain must identify the incoming sounds and make an analysis of those sounds and attach meaning to them. For instance, a person can have normal hearing but may have problems following directions or understanding conversations in noisy environments. They may have trouble sequencing a story or following a sequence of instructions. Because these individuals have normal hearing acuity, auditory processing problems may not be readily identified and may be mislabeled as a behavior problem or attention deficit disorder. In other words, auditory processing is what the listener does with the auditory information he/she receives.

There are certain basic aspects of auditory processing that play a role in the perception of speech. The frequency selectivity of the auditory system is described and used to derive the internal representation of the spectrum (the excitation pattern) of speech sounds. The perception of timbre and distinctions in quality between vowels are related to both static and dynamic aspects of the spectra of sounds. The perception of pitch and its role in speech perception, measures of the temporal resolution of the auditory system are important. The combined effects of frequency and temporal resolution give good insight into the internal representation of speech sounds.

Frequency Selectivity

The Concept of the Auditory Filter
The ability to resolve the complex sound into its constituent sinusoids is known as frequency selectivity. Frequency selectivity is important for the perception of loudness, pitch and timbre. Fletcher, (1940) Helmholtz (1863) have suggested that frequency selectivity can be modeled by considering the peripheral auditory system as a bank of band pass filters with overlapping pass bands. These are the auditory filters. According to Flectcher the basilar membrane within the cochlea provided the basis for the auditory filters. Each location on the basilar membrane responds to a limited range of frequencies, so each different point corresponds to a filter with a different centre frequency.

Frequency selectivity is best quantified by studying masking, a process by which the threshold of audibility for one sound is raised by the presence of another (masking) sound.

One method of measuring the shape of the auditory filter involves a procedure called a psychophysical tuning curve (PTC). To determine a PTC, the signal is fixed in level, usually at a very low level, say, 10dB above absolute threshold (10dB sensation levels, SL). The masker can be either a sinusoid or a narrowband noise. PTC indicates the masker level required to produce a fixed output power from the auditory filter as a function of frequency. Normally, a filter characteristic is determined by plotting the output from the filter for an input varying in frequency and fixed in level.

The second method is the notched-noise method. Patterson (1976) described a method of determining auditory filter shape which limits off-frequency listening and appears not to be influenced by beat detection. The signal is fixed in frequency, and the masker is a noise with a spectral notch centred at the signal frequency. The deviation of each edge of the notch from the centre frequency is denoted by Δf. The width of the notch is varied and the threshold of the signal is determined as a function of notch width. The analysis assumes that the filter is symmetric on a linear frequency scale as the notch is placed symmetrically around the signal frequency and asymmetries are not revealed.

As the width of the spectral notch is increased, less and less noise passes through the auditory filter. Thus the threshold of the signal drops. The amount of noise passing through the auditory filter is proportional to the area under the filter in the frequency range covered by the noise.

An alternative measure is the equivalent rectangular bandwidth (ERB), which is the bandwidth of a rectangular filter that has the same peak transmission as the filter of interest and passes the same total power for a white noise input. The ERB of the auditory filter is a little larger than the 3 dB bandwidth. In what follows, the mean ERB of the auditory filter determined using young listeners with normal hearing and using a moderate noise level is denoted ERB_N (where the subscript N denotes normal hearing). An equation describing the value of ERB_N as a function of centre frequency, F (in hertz), is (Glasberg & Moore 1990). A typical auditory filter shape determined using the notched-noise method. The filter is centred at 1 kHz. The relative response of the filter (in decibels) is plotted as a function of frequency.

Sometimes it is useful to plot psychoacoustical data on a frequency scale related to ERB_N, called the ERB_N-number scale. For example, the value of ERB_N for a centre frequency of 1 kHz is approximately 132 Hz, so an increase in frequency from 934 to 1066 Hz represents a step of one ERB_N-number. Each one-ERB_N step on the ERB_N-number scale corresponds approximately to a constant distance (0.9 mm) along the basilar membrane (Moore 1986). The ERB_N-number scale is conceptually similar to the Bark scale (Zwicker & Terhardt 1980), which has been widely used by speech researchers, although it differs somewhat in numerical values.

The notched-noise method has been extended to include conditions where the spectral notch in the noise is placed asymmetrically about the signal frequency. This allows the measurement of any asymmetry in the auditory filter, but the analysis of the results is more difficult, and has to take off-frequency listening into account (Patterson & Nimmo-Smith 1980). The results show that the auditory filter is reasonably symmetric at moderate sound levels, but becomes increasingly asymmetric at high levels, the low-frequency side becoming shallower than the high-frequency side. The filter shapes derived using the notched-noise method are quite similar to inverted PTCs (Glasberg *et al.* 1984), except that PTCs are slightly sharper around their tips, probably as a result of off-frequency listening and beat detection.

Masking Patterns and Excitation Patterns

In the masking experiments described so far, the frequency of the signal was held constant, while the masker was varied. These experiments are most appropriate for estimating the shape of the auditory filter at a given centre frequency. However, in many experiments the masker was held constant in both level and frequency, and the signal threshold was measured as a function of the signal frequency. The resulting functions are called masking patterns or masked audiograms.

Masking patterns show steep slopes on the low-frequency side (when the signal frequency is below that of the masker), of between 55 and 240dB per octave. The slopes on the high-frequency side are less steep and depend on the level of the masker.

The masking patterns do not reflect the use of a single auditory filter. Rather, for each signal frequency the listener uses a filter centered close to the signal frequency. Thus, the auditory filter is shifted as the signal frequency is altered. One way of interpreting the masking pattern is as a crude indicator of the excitation pattern of the masker (Zwicker & Fastl 1999). The excitation pattern is a representation of the effective amount of excitation produced by a stimulus as a function of characteristic frequency (CF) on the basilar membrane (Young 2008) and is plotted as effective level (in decibels) against CF. In the case of a masking sound, the excitation pattern can be thought of as representing the relative amount of vibration produced by the masker at different places along the basilar membrane. The signal is detected when the excitation it produces is some constant proportion of the excitation produced by the masker at places with CFs close to the signal frequency. Thus, the threshold of the signal as a function of frequency is proportional to the masker excitation level. The masking pattern should be parallel to the excitation pattern of the masker, but shifted vertically by a small amount. In practice, the situation is not so straightforward since the shape of the masking pattern is influenced by factors such as off-frequency listening, the detection of beats and combination tones (Moore *et al.* 1998; Alcántara *et al.* 2000) and by the physiological process of suppression (Delgutte 1990; see also Young 2008).

Moore & Glasberg (1983) have described a way of deriving the shapes of excitation patterns using the concept of the auditory filter. They suggested that the excitation pattern of a given sound can be thought of as the output of the auditory filters plotted as a function of their centre frequency. To calculate the excitation pattern of a sound, it is necessary to calculate the output of each auditory filter in response to that sound and to plot the output as a function of the filter centre frequency. The characteristics of the auditory filters are determined using the notched-noise method described earlier.

It should be noted that excitation patterns calculated do not take into account the physiological process of suppression, whereby the response to a given frequency component can be suppressed or reduced by a strong neighboring frequency component (Sachs & Kiang 1968; Young 2008). For speech sounds having spectra with strong peaks and valleys, such as vowels, suppression may have the effect of increasing the peak-to-valley ratio of the excitation pattern (Moore & Glasberg 1983). Also, the calculated excitation patterns are based on the power spectrum model of masking and do not take into account the effects of the relative phases of the components in complex sounds. However, it seems probable that excitation patterns provide a reasonable estimate of the extent to which the spectral features of complex sounds are represented in the auditory system.

Excitation Pattern of a Vowel Sound

There are some noteworthy observations about the excitation of vowels Firstly, the lowest few peaks in the excitation pattern do not correspond to formant frequencies, but rather to individual lower harmonics; these harmonics are resolved in the peripheral auditory system and can be heard out as separate tones under certain conditions (Plomp 1964a; Moore & Ohgushi 1993). Hence, the centre frequency of the first formant is not directly represented in the excitation pattern; if the frequency of the first formant is relevant for vowel identification (see Diehl 2008), then it must be inferred from the relative levels of the peaks corresponding to the individual lower harmonics. A second aspect of the excitation pattern is that, for this specific vowel, the second, third and fourth formants, which are clearly separately visible in the original spectrum, are not well resolved. Rather, they form a single prominence in the excitation pattern, with only minor ripples corresponding to the individual formants. Assuming that the excitation pattern does give a reasonable indication of the internal representation of the vowel, the perception of this vowel probably depends more on the overall prominence than on the frequencies of the individual formants. For other vowels, the higher formants often lead to separate peaks in the excitation pattern.

Frequency Selectivity in Cases of Impaired Hearing

In the developed countries, the most common cause of hearing loss is damage to the cochlea. This is usually associated with reduced frequency selectivity; the auditory filters are broader than normal (Moore 1998). As a result, the excitation patterns of complex sounds, such as vowels, are 'blurred' relative to those for normally hearing listeners. This makes it more difficult to distinguish the timbres of different vowel sounds. It also leads to increased susceptibility to masking by background sounds. For example, when trying to listen to a target talker in the presence of an interfering talker, a hearing-impaired person will be less able than normal to take advantage of differences in the short-term spectra of the two talkers, as described by Darwin (2008).

Across-Channel Processes in Masking

The discrimination and identification of complex sounds, including speech, require comparison of the outputs of different auditory filters.

Co Modulation Masking Release

Hall et al. (1984) were among the first to demonstrate that across-filter comparisons could enhance the detection of a sinusoidal signal in a fluctuating noise masker. The crucial feature for achieving this enhancement was that the fluctuations should be correlated across different frequency bands. One of their experiments was similar to a classic experiment of Fletcher (1940). The threshold for detecting a 1000 Hz, 400 ms sinusoidal signal was measured as a function of the bandwidth of a noise masker, keeping the spectrum level constant. The masker was centred at 1000 Hz. They used two types of masker. One was a random noise; this has irregular fluctuations in amplitude and the fluctuations are independent in different frequency regions. The other was a random noise which was modulated in amplitude at an irregular, low rate; a noise lowpass filtered at 50 Hz was used as a modulator. The modulation resulted in fluctuations in the amplitude of the noise which were the same in different frequency regions. This across-frequency correlation was called 'comodulation' by Hall et al. (1984).

The fact that the decrease in threshold with increasing bandwidth occurs only with the modulated noise indicating that fluctuations in the masker are critical and that these need to be correlated across frequency bands. Hence, this phenomenon has been called 'comodulation masking release' (CMR).It seems probable that across-filter comparisons of temporal envelopes are a general feature of auditory pattern analysis, which may play an important role in extracting signals from noisy backgrounds or separating competing sources of sound (Darwin 2008). As pointed out by Hall et al. (1984): 'Many real-life auditory stimuli have intensity peaks and valleys as a function of time in which intensity trajectories are highly correlated across frequency. This is true of speech, of interfering noise such as 'cafeteria' noise, and of many other kinds of environmental stimuli'. However, the importance of CMR for speech perception remains controversial. Some studies have suggested that it plays only a very minor role in the detection and identification of speech sounds in modulated background noise (Grose & Hall 1992; Festen 1993), although common modulation of target speech and background speech can lead to reduced intelligibility (Stone & Moore 2004). For synthetic speech in which the cues are impoverished compared to normal speech (sine-wave speech; see Remez *et al.* 1981), comodulation of the speech (amplitude modulation (AM) by a sinusoid) can markedly improve the intelligibility of the speech, both in quiet (Carrell & Opie 1992) and in background noise (Carrell 1993). The AM may help because it leads to perceptual fusion of the components of the sine-wave speech, so as to form an auditory object (Darwin 2008).

Profile Analysis

Green and colleagues (Green 1988) have carried out a series of experiments demonstrating that, even for stimuli without distinct envelope fluctuations, subjects are able to compare the outputs of different auditory filters to enhance the detection of a signal. They investigated the ability to detect an increment in the level of one component in a complex sound relative to the level of the other components; the other components are called the 'background'. Usually the complex sound has been composed of a series of equal-amplitude sinusoidal components, uniformly spaced on a logarithmic frequency scale. To prevent subjects from performing the task by monitoring the magnitude of the output of the single auditory filter centred at the frequency of the incremented component, the overall level of the whole stimulus was varied randomly from one stimulus to the next, over a relatively large range (typically 40dB). This makes the magnitude of the output of any single filter an unreliable cue to the presence of the signal.

Subjects were able to detect changes in the relative level of the signal of only 1–2dB. Such small thresholds could not be obtained by monitoring the magnitude of the output of a single auditory filter. Green and colleagues have argued that subjects performed the task by detecting a change in the shape or profile of the spectrum of the sound; hence the name 'profile analysis'. In other words, subjects can compare the outputs of different auditory filters, and can detect when the output of one changes relative to that of others, even when the overall level is varied. This is equivalent to detecting changes in the shape of the excitation pattern.

Timbre Perception

Timber refers to quality of a sound. It is all other aspect of sound other than its frequency, intensity and duration. Timbre identification depends strongly on two things, waveform of the steady part of the tone, and the way the spectrum changes with time, particularly at the onset or attack. This ability is probably built on pattern matching, a process that is well documented with vision. Once we have learned to identify a particular timbre, recognition is possible even if the pitch is changed or if parts of the spectrum are filtered out. (We are good enough at this that we can tell the pitch of low sounds when played through a sound system that does not reproduce the fundamentals.)

Timbre is usually defined as 'that attribute of auditory sensation in terms of which a listener can judge that two sounds similarly presented and having the same loudness and pitch are dissimilar' (ANSI 1994). The distribution of energy over frequency is one of the major determinants of timbre. However, timbre depends upon more than just the frequency spectrum of the sound; fluctuations over time can play an important role.

Timbre is multidimensional; there is no single scale along which the timbres of different sounds can be compared or ordered. Thus, a way is needed for describing the spectrum of a sound which takes into account this multidimensional aspect and which can be related to the subjective timbre. For steady sounds, a crude first approach is to look at the overall distribution of spectral energy. The 'brightness' or 'sharpness' (von Bismarck 1974) of sounds seems to be related to the spectral centroid. However, a much more quantitative approach has been described by Plomp and colleagues (Plomp 1970, 1976). They showed that the perceptual differences between different steady sounds, such as vowels, were closely related to the differences in the spectra of the sounds, when the spectra were specified as the levels in 18 1/3-octave frequency bands. A bandwidth of 1/3 octave is slightly greater than the ERB_N of the auditory filter over most of the audible frequency range. Thus, timbre is related to the relative level produced at the output of each auditory filter. In other words, the timbre of a steady sound is related to the excitation pattern of that sound.

It is probable that the number of dimensions required to characterize the timbre of steady sounds is limited by the number of ERB_{NS} required to cover the audible frequency range. This would give a maximum of approximately 37 dimensions. For a restricted class of sounds, such as vowels, a much smaller number of dimensions may be involved. It appears to be generally true, both for speech and non-speech sounds, that the timbres of steady tones are determined primarily by their magnitude spectra, although the relative phases of the components may also play a small role (Plomp & Steeneken 1969; Patterson 1987).

Differences in spectral shape are not always sufficient to allow the absolute identification of an 'auditory object', such as a musical instrument or a speech sound. One reason for this is that the magnitude and phase spectrum of the sound may be markedly altered by the transmission path and room reflections (Watkins 1991). In practice, the recognition of a particular timbre, and hence of an auditory object, may depend upon several other factors. Schouten (1968) has suggested that these include (i) whether the sound is periodic, having a tonal quality for repetition rates between approximately 20 and 20000 periods/s, or irregular, and having a noise-like character; (ii) whether the waveform envelope is constant, or fluctuates as a function of time, and in the latter case what the fluctuations are like; (iii) whether any other aspect of the sound (e.g. spectrum or periodicity) is changing as a function of time; and (iv) what the preceding and following sounds are like.

A powerful demonstration of the last factor may be obtained by listening to a stimulus with a particular spectral structure and then switching rapidly to a stimulus with a flat spectrum, such as white noise. A white noise heard in isolation may be described as 'colourless'; it has no pitch and has a neutral timbre. However, when a white noise follows immediately after a stimulus with spectral structure, the noise sounds 'coloured'. The coloration corresponds to the inverse of the spectrum of the preceding sound. For example, if the preceding sound is a noise with a spectral notch, the white noise has a pitch-like quality, with a pitch value corresponding to the centre frequency of the notch (Zwicker 1964). It sounds like a noise with a small spectral peak. A harmonic complex tone with a flat spectrum may be heard as having a vowel-like quality if it is preceded by a harmonic complex having a spectrum which is the inverse of that of a vowel (Summerfield *et al.* 1987).

The cause of this effect is not clear. Three types of explanation have been advanced, based on adaptation in the auditory periphery (see Young 2008), perceptual grouping (see Darwin 2008) and comparison of spectral shapes of the preceding and test sounds (Summerfield & Assmann 1987). All may play a role to some extent, depending on the exact properties of the stimuli. Whatever the underlying mechanism, it appears that the auditory system is especially sensitive to *changes* in spectral patterns over time (Kluender *et al.* 2003). This may be of value for communication in situations where the spectral shapes of sounds are (statically) altered by room reverberation or by a transmission channel with a non-flat frequency response.

The perceived timbre of brief segments of sounds can be strongly influenced by sounds that precede and follow those segments. In some cases, the observed effects appear to reflect relatively central perceptual compensation processes.

The Perception of Pitch

Pitch is usually defined as 'that attribute of auditory sensation in terms of which sounds can be ordered on a scale extending from low to high' (ANSI 1994). Variations in pitch give rise to a sense of melody. For speech sounds, variations in voice pitch over time convey intonation information, indicating whether an utterance is a question or a statement, and helping to identify stressed words. Voice pitch can also convey information about the sex, age and emotional state of the speaker (Rosen & Fourcin 1986). In some languages ('tone' languages), pitch and variations in pitch distinguish different lexical items.

Pitch is related to the repetition rate of the waveform of a sound; for a pure tone this corresponds to the frequency and for a periodic complex tone to the fundamental frequency, F0. There are, however, exceptions to this simple rule. Since voiced speech sounds are complex tones, the perception of pitch for complex tones is explained.

The Phenomenon of the Missing Fundamental

Although the pitch of a complex tone usually corresponds to its F0, the component with frequency equal to F0 does not have to be present for the pitch to be heard. Consider, as an example, a sound consisting of short impulses (clicks) occurring 200 times/s. This sound has a low pitch, which is very close to the pitch of a 200Hz sinusoid, and a sharp timbre. It contains harmonics with frequencies 200, 400, 600, 800, etc. Hz. However, if the sound is filtered so as to remove the 200 Hz component, the pitch does not alter; the only result is a slight change in the timbre of the note. All except a small group of mid-frequency harmonics can be eliminated, and the low pitch still remains, although the timbre becomes markedly different.

Schouten (1970) called the low pitch associated with a group of high harmonics the 'residue'. He pointed out that the residue is distinguishable, subjectively, from a fundamental component which is physically presented or from a fundamental which may be generated (at high sound pressure levels) by nonlinear distortion in the ear. Thus, the perception of a residue pitch does not require activity at the point on the basilar membrane which would respond maximally to a pure tone of similar pitch. Several other names have been used to describe residue pitch, including 'periodicity pitch', 'virtual pitch' and 'low pitch'. This paper will use the term low pitch. Even when the fundamental component of a complex tone is present, the low pitch of the tone is usually determined by harmonics other than the fundamental. Thus, the perception of a low pitch should not be regarded as unusual. Rather, low pitches are normally heard when listening to complex tones, including speech. For example, when listening over the telephone, the fundamental component for male speakers is usually inaudible, but the pitch of the voice can still be easily heard.

The Principle of Dominance

Ritsma (1967) carried out an experiment to determine which components in a complex sound are most important in determining its pitch. He presented complex tones in which the frequencies of a small group of harmonics were multiples of an F0 which was slightly higher or lower than the F0 of the remainder. The subject's pitch judgements were used to determine whether the pitch of the complex as a whole was affected by the shift in the group of harmonics. Ritsma found that: 'For fundamental frequencies in the range 100–400 Hz, and for SLs up to at least 50 dB above threshold of the entire signal, the frequency band consisting of the third, fourth and fifth harmonics tends to dominate the pitch sensation as long as its amplitude exceeds a minimum absolute level of about 10 dB above threshold'.

This finding has been broadly confirmed in other ways (Plomp 1967), although the data of Moore *et al.* (1984, 1985) show that there are large individual differences in which harmonics are dominant, and for some subjects the first two harmonics play an important role. Other data also show that the dominant region is not fixed in terms of harmonic number, but depends somewhat on absolute frequency (Plomp 1967; Patterson & Wightman 1976). For high F0s (above approx. 1000 Hz), the fundamental is usually the dominant component, while for very low F0s, approximately 50 Hz, harmonics above the fifth may be dominant (Moore & Glasberg 1988; Moore & Peters 1992). Finally, the dominant region shifts somewhat towards higher harmonics with decreasing duration (Gockel *et al.* 2005). For speech sounds, the dominant harmonics usually lie around the frequency of the first formant.

Discrimination of the Pitch of Complex Tones

When the F0 of a periodic complex tone changes, all of the components change in frequency by the same ratio, and a change in pitch is heard. The ability to detect such changes is better than the ability to detect changes in a sinusoid at F0 (Flanagan & Saslow 1958) and can be better than the ability to detect changes in the frequency of any of the sinusoidal components in the complex tone (Moore *et al.* 1984). This indicates that information from the different harmonics is combined or integrated in the determination of low pitch. This can lead to very fine discrimination; changes in F0 of approximately 0.2% can often be detected for F0s in the range 100–400 Hz.

The discrimination of F0 is usually best when low harmonics are present (Hoekstra & Ritsma 1977; Moore & Glasberg 1988; Shackleton & Carlyon 1994). Somewhat less good discrimination (typically 1–4%) is possible when only high harmonics are present (Houtsma & Smurzynski 1990). F0 discrimination can be impaired (typically by about a factor of two) when the two sounds to be discriminated also differ in timbre (Moore & Glasberg 1990); this can be the situation with speech sounds, where changes in F0 are usually accompanied by changes in timbre.

In speech, intonation is typically conveyed by differences in the pattern of F0 change over time. When the stimuli are dynamically varying, the ability to detect F0 changes is markedly poorer than when the stimuli are steady. Klatt (1973) measured thresholds for detecting differences in F0 for an unchanging vowel (i.e. one with static formant frequencies) with a flat F0 contour, and also for a series of linear glides in F0 around an F0 of 120 Hz. For the flat contour, the threshold was approximately 0.3 Hz. When both the contours were falling at the same rate (30 Hz over the 250 ms duration of the stimulus), the threshold increased markedly to 2 Hz. When the steady vowel was replaced by the sound

/ya/, whose formants change over time, thresholds increased further by 25–65%.

Generally, the F0 changes that are linguistically relevant for conveying stress and intonation are much larger than the limits of F0 discrimination measured psychophysically using steady stimuli. This is another reflection of the fact that information in speech is conveyed using robust cues that do not severely tax the discrimination abilities of the auditory system. Again, however, this may not be true for people with impaired hearing, for whom F0 discrimination is often much worse than normal (Moore & Carlyon 2005).

Perception of Pitch in Speech

Data on the perception of F0 contours in a relatively natural speech context were presented by Pierrehumbert (1979). She started with a natural nonsense utterance 'ma-MA-ma-ma-MA-ma', in which the prosodic pattern was based on the sentence 'The baker made bagels'. The stressed syllables (MA) were associated with peaks in the F0 contour. She then modified the F0 of the second peak, over a range varying from below to above the F0 of the first peak.

Subjects listened to the modified utterances and were required to indicate whether the first or second peak was higher in pitch. The results reflected what she called 'normalization for expected declination'; when the two stressed syllables sounded equal in pitch, the second was actually lower in F0. For first peak values of 121 and 151 Hz, the second peak had to be shifted over a range of approximately 20 Hz to change judgements from 75% 'second peak lower' to 75% 'second peak higher'. This indicates markedly poorer discriminability than found for steady stimuli. Similarly, 't Hart (1981) found that about a 19% difference was necessary for successive pitch movements in the same direction to be reliably heard as different in extent.

Hermes & van Gestel (1991) studied the perception of the excursion size of prominence-lending F0 movements in utterances resynthesized in different F0 registers. The task of the subjects was to adjust the excursion size in a comparison stimulus in such a way that it lent equal prominence to the corresponding syllable in a fixed test stimulus. The comparison stimulus and the test stimulus had F0s running parallel on either a logarithmic frequency scale, an ERB_N-number scale, or a linear frequency scale. They found that stimuli were matched in such a way that the average excursion sizes in different registers were equal when the ERB_N-number scale was used. In other words, the perceived prominence of F0 movements is related to the size of those movements expressed on an ERB_N-number scale.

Temporal Analysis

Time is a very important dimension in hearing, since almost all sounds change over time. For speech, much of the information appears to be carried in the changes themselves, rather than in the parts of the sounds which are relatively stable (Kluender et al. 2003).

In characterizing temporal analysis, it is essential to take account of the filtering that takes place in the peripheral auditory system. Temporal analysis can be considered as resulting from two main processes: analysis of the time pattern occurring within each frequency channel and comparison of the time patterns across channels. This paper focuses on the first of these.

A major difficulty in measuring the temporal resolution of the auditory system is that changes in the time pattern of a sound are generally associated with changes in its magnitude spectrum—the distribution of energy over frequency. Thus, the detection of a change in time pattern can sometimes depend not on temporal resolution *per se,* but on the detection of the spectral change.

Sometimes, the detection of spectral changes can lead to what appears to be extraordinarily fine temporal resolution. For example, a single click can be distinguished from a pair of clicks when the gap between the two clicks in a pair is only a few tens of microseconds, an ability that depends upon spectral changes at very high frequencies (Leshowitz 1971). Although spectrally based detection of temporal changes can occur for speech sounds, this paper focuses on experimental situations which avoid the confounding effects of spectral cues.

There have been two general approaches to avoiding the use of cues based on spectral changes. One is to use signals whose magnitude spectrum is not changed when the time pattern is altered. For example, the magnitude spectrum of white noise remains flat if a gap is introduced into the noise. The second approach uses stimuli whose spectra are altered by the change in time pattern, but extra background sounds are added to mask the spectral changes. Both of these approaches will be considered.

Within-Channel Temporal Analysis Using Broadband Sounds

Temporal resolution can be measured by detecting threshold for a gap in a broadband noise. The gap threshold is typically 2–3ms (Plomp 1964).

At very low sound levels the threshold increases but is invariant with levels for moderate to high levels.

Ronken (1970) used pairs of clicks differing in amplitude as stimuli. One click (A) had amplitude greater than that of the other click(B). Typically, the amplitude of A click was twice that of B. Subjects were required to distinguish click pairs differing in the order of A and B: either AB or BA. The ability to do this was measured as a function of the time interval or gap between A and B. Ronken found that subjects could distinguish the click pairs for gaps down to 2–3ms. The limit to temporal resolution found in this task is similar to that found for the detection of a gap in broadband noise. It should be noted that, in this task, subjects do not hear the individual clicks within a click pair. Rather, each click pair is heard as a single sound with its own characteristic quality. For example, the two click pairs AB and BA might sound like 'tick' and 'tock'.

A more general approach is to measure the threshold for detecting changes in the amplitude of a sound as a function of the rapidity of the changes. In the simplest case, white noise is sinusoidally amplitude modulated, and the threshold for detecting the modulation is determined as a function of modulation rate. The function relating threshold to modulation rate is known as a temporal modulation transfer function (TMTF; Viemeister 1979). An example of a TMTF is shown in figure 10 (Bacon & Viemeister 1985). The thresholds are expressed as $20\log m$, where m is the modulation index ($m=0$ corresponds to no modulation and $m=1$ corresponds to 100% modulation). For low modulation rates, performance is limited by the amplitude resolution of the ear, rather than by temporal resolution. Thus, the threshold is independent of modulation rate for rates up to approximately 50Hz. As the rate increases beyond 50Hz, temporal resolution starts to have an effect; performance worsens, and for rates above approximately 1000Hz the modulation is hard to detect at all. Thus, sensitivity to modulation becomes progressively less as the rate of modulation increases. The shapes of TMTFs do not vary much with overall sound level, but the ability to detect the modulation worsens at low sound levels. Over the range of modulation rates important for speech perception, below approximately 50Hz (Steeneken & Houtgast 1980; Drullman et al. 1994 *a,b*), the sensitivity to modulation is rather good.

A temporal modulation transfer function (TMTF). A broadband white noise was sinusoidally amplitude modulated, and the threshold amount of modulation required for detection is plotted as a function of modulation rate. The amount of modulation is specified as $20\log m$, where m is the modulation index. The higher the sensitivity to modulation, the more negative is this quantity.

Within-Channel Temporal Analysis Using Narrowband Sounds

Temporal resolution of the auditory system varies with centre frequency is a question that cannot be answered with broadband sounds. Using narrowband stimuli that excite only one, or a small number, of auditory channels, the issue can be examined.

Green (1973) used stimuli where each stimulus consisted of a brief pulse of a sinusoid in which the level of the first half of the pulse was 10dB different from that of the second half. Subjects were required to distinguish two signals, differing in whether the half with the high level was first or second. Green measured performance as a function of the total duration of the stimuli. The threshold was similar for centre frequencies of 2 and 4kHz, and was between 1 and 2ms. However, the threshold was slightly higher for a centre frequency of 1kHz, being between 2 and 4ms.

Performance in this task was actually a non-monotonic function of duration. Performance was good for durations in the range 2–6ms, worsened for durations around 16ms, and then improved again as the duration was increased beyond 16ms. For the very short durations, subjects listened for a difference in quality between the two sounds—rather like the 'tick' and 'tock' described earlier for Ronken's stimuli. At durations around 16ms, the tonal quality of the bursts became more prominent and the quality differences were harder to hear. At much longer durations, the soft and loud segments could be separately heard, in a distinct order. It appears, therefore, that performance in this task was determined by two separate mechanisms, one based on timbre differences associated with the difference in time pattern, and the other based on the perception of a distinct succession of auditory events.

Several researchers have measured thresholds for detecting gaps in narrowband sounds, either noises (Fitzgibbons 1983; Shailer & Moore 1983; Buus & Florentine 1985; Eddins et al. 1992) or sinusoids (Shailer & Moore 1987; Moore et al. 1993). When a temporal gap is introduced into a narrowband sound, the spectrum of the sound is altered. Energy 'splatter' occurs outside the nominal frequency range of the sound. To prevent the splatter being detected, the sounds are presented in a background sound, usually a noise, designed to mask the splatter.

Gap thresholds for noise bands decrease with increasing bandwidth but show little effect of centre frequency when the bandwidth is held constant. For noises of moderate bandwidth (a few hundred hertz), the gap threshold is typically approximately 10ms. Gap thresholds for narrowband noises tend to decrease with increasing sound level for levels up to approximately 30dB above absolute threshold, but remain roughly constant after that.

Shailer & Moore (1987) showed that the detectability of a gap in a sine wave was strongly affected by the phase at which the sinusoid was turned off and on to produce the gap (Shailer & Moore 1987). Only the simplest case is considered here, called 'preserved phase' by Shailer & Moore (1987)

In this case, the sinusoid was turned off at a positive-going zero crossing (i.e. as the waveform was about to change from negative to positive values) and it started (at the end of the gap) at the phase it would have had if it had continued without interruption. Thus, for the preserved-phase condition it was as if the gap had been 'cut out' from a continuous sinusoid. For this condition, the detectability of the gap increased monotonically with increasing gap duration.

Shailer & Moore (1987) found that the threshold for detecting a gap in a sine wave was roughly constant at approximately 5ms for centre frequencies of 400, 1000 and 2000Hz. Moore et al. (1993) found that gap thresholds were almost constant at 6–8ms over the frequency range 400–2000Hz, but increased somewhat at 200Hz, and increased markedly, to approximately 18ms, at 100Hz. Individual variability also increased markedly at 100Hz.

Overall, the results of experiments using narrowband stimuli indicate that temporal resolution does not vary markedly with centre frequency, except perhaps for a worsening at very low frequencies (200Hz and below). Gap thresholds for narrowband stimuli are typically higher than those for broadband noise. However, for moderate noise bandwidths, gap thresholds are typically approximately 10ms or less. The smallest detectable gap is usually markedly larger than temporal gaps that are relevant for speech perception (for example, 'sa' and 'sta' may be distinguished by a temporal gap lasting several tens of milliseconds).

REFERENCES

Aibara, R., Welsh, J. T., Puria, S., & Goode R. L. (2001). Human middle-ear sound transfer function and cochlear input impedance. *Hearing Research, 152,* 100–109.

Alcantara, J. I., Moore, B. C. J., & Vickers, D.A. (2000). The relative role of beats and combination tones in determining the shapes of masking patterns at 2 kHz: I. Normal-hearing listeners. *Hearing Research, 148,* 63–73.

Anagnostakos, N. P. (1978). *Principles of Anatomy and Physiology* (2nd ed.). San: Francisco Canfield.

Armstrong, K. F. (1969). *Anatomy and Physiology for Nurses.* London: ELBS.

Bacon, S. P., & Viemeister, N. F (1985). Temporal modulation transfer functions in normal-hearing and hearing-impaired subjects. *Audiology, 24,* 117–134.

Brungart, D. S., Simpson, B. D., Darwin, C. J., Arbogast, T. L., & Kidd, G., (2005). Across-ear interference from parametrically degraded synthetic speech signals in a dichotic cocktail-party listening task. *Journal of Acoustical Society of America,117,* 292–304.

Buus, S., & Florentine, M. (1985). Gap detection in normal and impaired listeners: the effect of level and frequency. In: Michelsen A, editor. Time resolution in auditory systems. Springer; New York: 1985. pp. 159–179.

Bateman, H. (1977). *A Clinical Approach to Speech Anatomy and Physiology.* Springfield, USA: Charles C Thomas.

Brimble. (n.d.). *Physiology Anatomy and Health.* New York: Macmillan Company.

Carrell, T. (1993). The effect of amplitude comodulation on extracting sentences from noise: evidence from a variety of contexts. *Journal of Acoustical Society of America,93,* 2327.

Carrell, T. D, Opie, J. M. (1992).The effect of amplitude comodulation on auditory object formation in sentence perception. *Perception and Psychophysics, 52,* 437–445.

Christensen, J. B. (1974). *Dynamic anatomy& Physiology* (Vol. 4th). New York: McGraw- Mill Book Company.

Clark, W. W. (2008). *Anatomy & Physiology of Hearing for Audiologists.* New York: Thomas Delmar Learning.

Culbertson, W. R. (2006). *Anatomy & Physiology Study Guide for Speech & Hearing.* San Diego: Plural Publishing.

Culbertson, W. R. (2013). *Anatomy & Physiology Study Guide for Speech & Hearing* (Vol. 2nd edition). San Diego: Plural Publishers.

Cf, D. (n.d.). *Introduction to the Anatomy and Physiology of speech Mechanism.* Springfield, USA: Charles C Thomas Publishers.

Darwin, C. J. (2008). Listening to speech in the presence of other sounds. *Philosophical Transactions of the Royal Society B, 363,* 1011–1021.

Dau, T., Kollmeier, B., & Kohlrausch, A. (1997). Modeling auditory processing of amplitude modulation I. Detection and masking with narrowband carriers. *Journal of Acoustical Society of America,* 102, 2892–2905.

Dau, T., Kollmeier, B., & Kohlrausch, A. (1997).Modeling auditory processing of amplitude modulation II. Spectral and temporal integration. *Journal of Acoustical Society of America, 102,* 2906–2919.

Delgutte, B. (1990). Physiological mechanisms of psychophysical masking: observations from auditory-nerve fibers. *Journal of Acoustical Society of America, 87,* 791–809.

Diehl, R. L. (2008). Acoustic and auditory phonetics: the adaptive design of speech sound systems. *Philosophical Transactions of the Royal Society B, 363,* 965–978.

Drullman, R., Festen, J.M., & Plomp R. (1994).Effect of reducing slow temporal modulations on speech reception. *Journal of Acoustical Society of America,95,* 2670–2680.

Drullman, R., Festen, J. M., & Plomp, R. (1994).Effect of temporal envelope smearing on speech reception. *Journal of Acoustical Society of America,95,* 1053–1064.

Eddins, D. A., Hall, J. W., & Grose, J. H. (1992). Detection of temporal gaps as a function of frequency region and absolute noise bandwidth. *Journal of Acoustical Society of America,91,* 1069–1077.

Egan, J. P., & Hake, H. W. (1950).On the masking pattern of a simple auditory stimulus. *Journal of Acoustical Society of America, 22,* 622–630.

Festen, J. M. (1993). Contributions of comodulation masking release and temporal resolution to the speech–reception threshold masked by an interfering voice. *Journal of Acoustical Society of America,94,* 1295–1300.

Fitzgibbons, P. J. (1983). Temporal gap detection in noise as a function of frequency, bandwidth and level. *Journal of Acoustical Society of America, 74,* 67–72.

Flanagan, J. L, & Saslow, M. G. (1958). Pitch discrimination for synthetic vowels. *Journal of Acoustical Society of America,30,* 435–442.

Fletcher, H. (1940). Auditory patterns. *Reviews of Modern Physics, 12,* 47–65.

Fourcin, A. J., & Abberton, E. (1977). Laryngograph studies of vocal-fold vibration. *Phonetica, 34,* 313–315.

Drumright, D. G. (1997). *Anatomy & Physiology.* san diego, london: Singular publishing Group.

Grant, A. W. (1998). *Ross ans Wilson anatomy and Physiology in Health and Illness.* London: churchill livingstone.

Greisheimer, E. M. (1945). *Physiology and Anatomy* (Vol. 5th). Philadelphia: Lippincott company.

Grollman, S. (1969). *Laboratory Manual of Mammalian Anatomy and Physiology* (Vol. 2nd). New York: Macmillan Company.

Glasberg, B. R., Moore, B. C. J. (1990). Derivation of auditory filter shapes from notched-noise data. *Hearing Research, 47,* 103–138.

Glasberg, B. R, & Moore, B. C. J. (2000). Frequency selectivity as a function of level and frequency measured with uniformly exciting notched noise. *Journal of Acoustical society of America,108,* 2318–2328.

Glasberg, B. R., & Moore, B. C. J. (2002).A model of loudness applicable to time-varying sounds. *Journal of Audio Engineering Society,50,* 331–342.

Glasberg, B. R., Moore, B. C. J., Patterson, R. D., & Nimmo-Smith, I (1984).Dynamic range and asymmetry of the auditory filter. *Journal of Acoustical Society of America, 76,* 419–427.

Gockel, H., Carlyon, R. P., & Plack, C. J. (2005). Dominance region for pitch: effects of duration and dichotic presentation. *Journal of Acoustical Society of America,117,* 1326–1336.

Green, D. M. (1973).Temporal acuity as a function of frequency. *Journal of Acoustical Society of America,54,* 373–379.

Green, D.M. (1988). Oxford University Press; Oxford, UK: Profile analysis.

Grose, J. H., & Hall, J. W. (1992).Comodulation masking release for speech stimuli. *Journal of Acoustical Society of America,91,* 1042–1050.

Hall, J. W., Haggard, M. P., & Fernandes, M. A. (1984).Detection in noise by spectro-temporal pattern analysis. *Journal of Acoustical Society of America,76,* 50–56.

Hall, J.W, Grose, J. H., & Mendoza, L. (1995). Across-channel processes in masking. In: Moore, B. C. J, editor. Hearing Academic Press; San Diego, CA: pp. 243–266.

Hermes, D. J, & van Gestel, J. C. (1991).The frequency scale of speech intonation. *Journal of Acoustical Society of America,90,* 97–102.

Hoekstra, A., & Ritsma, R. J. (1977).Perceptive hearing loss and frequency selectivity. In: Evans E.F, Wilson J.P, editors. Psychophysics and physiology of hearing. Academic; London, UK: pp. 263–271.

Houtsma, A. J. M, & Smurzynski, J. (1990). Pitch identification and discrimination for complex tones with many harmonics. *Journal of Acoustical Society of America,87,* 304–310.

Irino, T., & Patterson, R. D. (2001).A compressive gamma chirp auditory filter for both physiological and psychophysical data. *Journal of Acoustical Society of America,109,* 2008–2022.

Johnson-Davies, D., & Patterson, R. D. (1979). Psychophysical tuning curves: restricting the listening band to the signal region. *Journal of Acoustical Society of America, 65,* 765–770.

Jayaveera, K. N. (2010). *Human Anatomy Physiology and Health Education* (Vol. 1st). New Delhi: S. Chand and Company.

Kay, R. H., & Mathews, D. R. (1972). On the existence in human auditory pathways of channels selectively tuned to the modulation present in frequency-modulated tones. *Journal of Physiology, 225,* 657–677.

Klatt, D. H. (1973). Discrimination of fundamental frequency contours in speech: implications for models of pitch perception. *Journal of Acoustical Society of America,53,* 8–16.

Kluender, K. R, Coady, J. A, & Kiefte, M. (2003).Sensitivity to change in perception of speech. *Speech Communication. 41,* 59–69.

Kluk, K., & Moore, B. C. J. (2004).Factors affecting psychophysical tuning curves for normally hearing subjects. *Hearing Research, 194,* 118–134.

Kohlrausch, A., Fassel, R., & Dau, T. (2000).The influence of carrier level and frequency on modulation and beat-detection thresholds for sinusoidal carriers. *Journal of Acoustical Society of America,108,* 723–734.

Kortekaas, R.W., & Kohlrausch, A. (1999).Psychoacoustical evaluation of PSOLA II. Double-formant stimuli and the role of vocal perturbation. *Journal of Acoustical Society of America, 105,* 522–535.

Langner, G., & Schreiner, C. E. (1988).Periodicity coding in the inferior colliculus of the cat. I. Neuronal mechanisms. *Journal of Neurophysiology, 60,* 1799–1822.

Leshowitz, B. (1971). Measurement of the two-click threshold. *Journal of acoustical society of America, 49,* 426–466.

Licklider, J. C., & Pollack, I. (1948).Effects of differentiation, integration and infinite peak clipping upon the intelligibility of speech. *Journal of Acoustical Society of America, 20,* 42–52.

Lopez-Poveda, E. A., & Meddis, R. (2001).A human nonlinear cochlear filterbank. *Journal of Acoustical Society of America,* 110, 3107–3118.

Moore, B. C. J. (1986). Parallels between frequency selectivity measured psychophysically and in cochlear mechanics. *Scandinavian Audiology, 25 (Suppl.),* 139–152.

Moore, B. C. J. (1992). Across-channel processes in auditory masking. *Journal of Acoustical Society of America,13,* 25–37.

Moore B. C. J. (1996). Masking in the human auditory system. In: Gilchrist N, Grewin C, editors. Collected papers on digital audio bit-rate reduction. Audio Engineering Society; New York, NY: pp. 9–19.

Moore, B. C. J (1998). *Cochlear hearing loss.* London, UK, Whurr Publishers Ltd.

Moore, B. C. J. (2003). An introduction to the psychology of hearing (5th edn). Academic Press; San Diego, CA:

Moore, B. C. J. (2003). Speech processing for the hearing-impaired: successes, failures, and implications for speech mechanisms. *Speech Communication, 41,* 81–91.

Moore, B. C. J. (2003). Temporal integration and context effects in hearing. *Journal of Phonetics.31,* 563–574.

Moore, B. C. J., & Carlyon, R. P. (2005).Perception of pitch by people with cochlear hearing loss and by cochlear implant users. In: Plack C.J, Oxenham A.J, Fay R.R, Popper A.N, editors. Pitch perception. Springer; New York, NY: pp. 234–277.

Moore, B. C. J, & Glasberg, B. R. (1983). Masking patterns of synthetic vowels in simultaneous and forward masking. *Journal of Acoustical Society of America, 73,* 906–917.

Moore, B. C. J, & Glasberg, B. R. (1893).Suggested formulae for calculating auditory-filter bandwidths and excitation patterns. *Journal of Acoustical Society of America,74,* 750–753.

Moore B. C. J & Glasberg, B. R. (1987).Formulae describing frequency selectivity as a function of frequency and level and their use in calculating excitation patterns. *Hearing Research, 28,* 209–225.

Moore, B. C. J, & Glasberg, B. R. (1988).Effects of the relative phase of the components on the pitch discrimination of complex tones by subjects with unilateral and bilateral cochlear impairments. In: Duifhuis H, Wit H, Horst J, editors. Basic issues in hearing. Academic Press; London, UK: pp. 421–430.

Moore, B. C. J, & Glasberg, B. R. (1990).Frequency discrimination of complex tones with overlapping and non-overlapping harmonics. *Journal of Acoustical Society of America,87,* 2163–2177.

Moore, B. C. J, & Ohgushi, K. (1993).Audibility of partials in inharmonic complex tones. *Journal of Acoustical Society of America,93,* 452–461.

Moore, B. C. J, & Peters, R. W. (1993). Pitch discrimination and phase sensitivity in young and elderly subjects and its relationship to frequency selectivity. *Journal of Acoustical Society of America, 91,* 2881–2893.

Moore, B. C. J., & Shailer, M. J. (1992).Modulation discrimination interference and auditory grouping. *Philosophical Transactions of the Royal Society B.336,* 339–346.

Moore, B. C. J, Glasberg, B. R, & Shailer, M. J. (1984). Frequency and intensity difference limens for harmonics within complex tones. *Journal of Acoustical Society of America. 75,* 550–561.

Moore, B. C. J., Glasberg, B. R., & Peters, R. W (1985).Relative dominance of individual partials in determining the pitch of complex tones. *Journal of acoustical society of America, 77,* 1853–1860.

Moore, B. C. J., Glasberg, B. R., Plack, C. J., & Biswas, A. K. (1988). The shape of the ear's temporal window. *Journal of Acoustical Society of America,83,* 1102–1116.

Moore, B. C. J., Glasberg, B. R., Gaunt, T., & Child, T. (1991).Across-channel masking of changes in modulation depth for amplitude-and frequency-modulated signals. *Quarterly Journal of Experimental Psychology, 43A,* 327–347.

Moore, B. C. J., Peters, R. W., & Glasberg, B. R. (1993). Detection of temporal gaps in sinusoids: effects of frequency and level. *Journal of Acoustical Society of America, 93,* 1563–1570.

Moore, B. C. J., Alcantara, J, I., & Dau, T. (1998) Masking patterns for sinusoidal and narrowband noise maskers. *Journal of Acoustical Society of America, 104,* 1023–1038.

Moore, B. C. J, Glasberg, B. R., Flanagan, H. J., & Adams, J. (2006).Frequency discrimination of complex tones; assessing the role of component resolvability and temporal fine structure. *Journal of Acoustical Society of America, 119,* 480–490.

Musiek, F. E. (2007). *Auditory System: Anatomy, Physiology & Clinical Correlates.* Boston: Pearson Education.

Mary, P. (2000). *Sarups dictionary of Anatomy and Physiology* (1st ed.). New Delhi: Sarup and sons.

Moller, A. R. (2013). *Hearing: Anatomy, Physioogy and Disorders of the Auditory system* (Vol. 3rd). San Diego: Plural Publishers.

Negus, V. E. (1962). *The Comparitive Anatomy and physiology of larynx.* New York: Hafner Publications.

O'Loughlin, B. J, & Moore, B. C. J. (1981). Improving psychoacoustical tuning curves. *Hearing Research, 5,* 343–346.

O'Loughlin B. J, & Moore, B. C. J. (1981). Off-frequency listening: effects on psychoacoustical tuning curves obtained in simultaneous and forward masking. *Journal of Acoustical Society of America,69,* 1119–1125.

Oxenham, A. J, & Moore, B. C. J. (1994). Modeling the additivity of nonsimultaneous masking. *Hearing Research, 80,* 105–118.

Patterson, R. D. (1976). Auditory filter shapes derived with noise stimuli. *Journal of Acoustical Society of America,59,* 640–654.

Patterson, R. D., & Moore, B. C. J. (1986).Auditory filters and excitation patterns as representations of frequency resolution. In: Moore B.C.J, editor. Frequency selectivity in hearing. Academic; London, UK: pp. 123–177.

Patterson, R. D., & Nimmo-Smith, I. (1980).Off-frequency listening and auditory filter asymmetry. *Journal of Acoustical Society of America,67*: 229–245.

Patterson, R. D, & Wightman, F. L. (1976).Residue pitch as a function of component spacing. *Journal of Acoustical Society of America, 59,* 1450–1459.

Pierrehumbert, J. (1979). The perception of fundamental frequency declination. *Journal of Acoustical Society of America,66,* 363–368.

Plack, C. J, & Moore, B. C. J. (1990). Temporal window shape as a function of frequency and level. *Journal of Acoustical Society of America, 87,* 2178–2187.

Plomp, R. (1964). The ear as a frequency analyzer. *Journal of Acoustical Society of America,36,* 1628–1636.

Plomp, R. (1964). The rate of decay of auditory sensation. *Journal of Acoustical Society of America,36,* 277–282.

Plomp, R. (1967). Pitch of complex tones. *Journal of Acoustical Society of America,41,* 1526–1533.

Plomp, R. (1970). Timbre as a multidimensional attribute of complex tones. In: Plomp R, Smoorenburg G.F, editors. Frequency analysis and periodicity detection in hearing. Sijthoff; Leiden, The Netherlands:

Plomp, R. (1976). *Aspects of tone sensation.* Academic Press; London, UK.

Plomp, R., & Steeneken, H. J. M. (1969).Effect of phase on the timbre of complex tones. *Journal of Acoustical Society of America, 46,* 409–421.

Pollack, I. (1968). Periodicity discrimination for auditory pulse trains. *Journal of Acoustical Society of America,43,* 1113–1119.

Remez, R. E., Rubin, P. E., Pisoni, D. B., & Carrell, T. D. (1981). Speech perception without traditional speech cues. *Science, 212,* 947–950.

Ritsma, R. J. (1967). Frequencies dominant in the perception of the pitch of complex sounds. *Journal of Acoustical Society of America, 42,* 191–198.

Ronken, D. (1970). Monaural detection of a phase difference between clicks. *Journal of Acoustical Society of America,47,* 1091–1099.

Rosen, S., Fourcin, A. (1986).Frequency selectivity and the perception of speech. In: Moore B.C.J, editor. Frequency selectivity in hearing. Academic; London, UK: pp. 373–487.

Rosen, S., Baker, R. J, & Darling, A. (1998). Auditory filter nonlinearity at 2kHz in normal hearing listeners. *Journal of Acoustical Society of America, 103,* 2539–2550.

Ruggero, M. A., Rich, N. C., Recio, A., Narayan, S. S., & Robles, L. (1997).Basilar-membrane responses to tones at the base of the chinchilla cochlea. *Journal of Acoustical Society of America,101,* 2151–2163.

Ross, J. S. (1982). *Foundations of Anatomy and Physiology* (Vol. 5th). London: ELBS.

Sears., G. W. (1965). *Anatomy and Physiology for Nurses.* london: Edward Ernold.

Seeley, R. S. (2002). *Essentials of Anatomy & Physiology* (Vol. 4th). New York: McGraw- Mill Book Company.

Seikel, J. A. (2004). *Essentials of Anatomy Physiology for Communication Disorders.* New York: Thomas Delmar Learing.

Seikel, J. A. (2010). *Anatomy and Physiology for Speech Language and Hearing* (Vol. 4th). NY: Delmar.

Sheets, B. V. (1973). *Anatomy anf Physiology of Speech Mechanism.* New York: Bobbs-Merill.

Sylvester, P. E. (1964). *Applied Anatomy and Physiology for Nurses.* Oxford: Blockwell.

Saberi, K., & Perrott, D. R. (1999).Cognitive restoration of reversed speech. *Nature, 398,* 760.

Sachs., M. B, & Kiang., N. Y. S. (1968). Two-tone inhibition in auditory nerve fibers *Journal of Acoustical Society of America, 43,* 1120–1128.

Schouten, J.F. (1970). The residue revisited. In: Plomp R, Smoorenburg G.F, editors. Frequency analysis and periodicity detection in hearing. Sijthoff; Leiden, The Netherlands: pp. 41–54.

Schreiner, C. E, Urbas, J. V. (1986). Representation of amplitude modulation in the auditory cortex of the cat I. The anterior auditory field (AAF). *Hearing Research, 21,* 227–241.

Shackleton, T. M., & Carlyon, R. P. (1994).The role of resolved and unresolved harmonics in pitch perception and frequency modulation discrimination. *Journal of Acoustical Society of America,95,* 3529–3540.

Shailer, M. J., & Moore, B. C. J. (1983).Gap detection as a function of frequency, bandwidth and level. *Journal of Acoustical Society of America, 74,* 467–473.

Shailer, M. J., & Moore, B. C. J. (1987). Gap detection and the auditory filter: phase effects using sinusoidal stimuli. *Journal of Acoustical Society of America, 81,* 1110–1117.

Shannon, R.V., Zeng, F. G., Kamath, V., Wygonski, J., & Ekelid, M. (1995).Speech recognition with primarily temporal cues. *Science, 270,* 303–304.

Shaw, E. A. G. (1974). Transformation of sound pressure level from the free field to the eardrum in the horizontal plane. *Journal of Acoustical Society of America,56,* 1848–1861.

Steeneken, H. J. M., & Houtgast, T. (1980).A physical method for measuring speech-transmission quality. *Journal of Acoustical Society of America, 69,* 318–326.

Stone, M. A, & Moore, B. C. J. (2004).Side effects of fast-acting dynamic range compression that affect intelligibility in a competing speech task. *Journal of Acoustical Society of America, 116,* 2311–2323.

Summerfield, A. Q., Assmann, P. (1987). Auditory enhancement in speech perception. In: Schouten M.E.H, editor. The psychophysics of speech perception. Martinus Nijhoff; Dordrecht, The Netherlands: pp. 140–150.

Summerfield, A. Q., Sidwell, A. S., & Nelson, T. (1987).Auditory enhancement of changes in spectral amplitude. *Journal of Acoustical Society of America,* 81, 700–708.

Sumner, C. J., Lopez-Poveda, E. A., O'Mard, L. P., Meddis, R. (2002).A revised model of the inner-hair cell and auditory-nerve complex. *Journal of Acoustical Society of America, 111,* 2178–2188.

Verhey, J. L., Dau, T., & Kollmeier, B. (1999). Within-channel cues in comodulation masking release (CMR): experiments and model predictions using a modulation-filterbank model. *Journal of Acoustical society of America, 106,* 2733–2745.

Viemeister, N. F. (1979). Temporal modulation transfer functions based on modulation thresholds. *Journal of Acoustical society of America, 66,* 1364–1380.

Vogten, L. L. (1978). Low-level pure-tone masking: a comparison of "tuning curves" obtained with simultaneous and forward masking. *Journal of Acoustical Society of America,63,* 1520–1527.

von Bismarck, G. (1974). Sharpness as an attribute of the timbre of steady sounds.*Acustica.30,* 159–172.

Watkins, A. J. (1991). Central, auditory mechanisms of perceptual compensation for spectral-envelope distortion. *Journal of Acoustical Society of America, 90,* 2942–2955.

Watkins, A. J, & Makin, S. J. (1996). Effects of spectral contrast on perceptual compensation for spectral-envelope distortion. *Journal of Acoustical Society of America, 99,* 3749–3757.

Watkins, A. J., & Makin, S. J. (1996). Some effects of filtered contexts on the perception of vowels and fricatives. *Journal of Acoustical Society of America,99,* 588–594.

Yost, W. A., & Sheft, S. (1989). Across-critical-band processing of amplitude-modulated tones. *Journal of Acoustical Society of America, 85,* 848–857.

Yost, W. A., Sheft, S., & Opie, J. (1989).Modulation interference in detection and discrimination of amplitude modulation. *Journal of Acoustical Society of America,86,* 2138–2147.

Young, E. D. (2008). Neural representation of spectral and temporal information in speech. *Philosophical Transactions of the Royal Society B. 363,* 923–945.

Yumoto, E., & Gould, W. J., & Baer, T. (1982).Harmonics-to-noise ratio as an index of the degree of hoarseness. *Journal of Acoustical Society of America, 71:*1544–1549.

Zhang, X., Heinz, M. G., Bruce, I. C., & Carney, L. H. (2001). A phenomenological model for the responses of auditory-nerve fibers: I. Nonlinear tuning with compression and suppression. *Journal of Acoustical Society of America,109,* 648–670.

Zwicker, E. (1964). 'Negative afterimage' in hearing. *Journal of Acoustical Society of America,36,* 2413–2415.

Zwicker, E., & Fastl, H. (1999).*Psychoacoustics–facts and models (*2nd edn).Springer; Berlin, Germans.

Zwicker, E., & Terhardt, E. (1980).Analytical expressions for critical band rate and critical bandwidth as a function of frequency. *Journal of Acoustical Society of America,68,* 1523–1525.

Chapter 4

INFLUENCE OF MUSIC ON OTHER DOMAINS

MUSIC AND STRUCTURAL CHANGES IN THE BRAIN

In the human brain, one of the most powerful sources of auditory stimulation is provided by music (Sacks, 2006). Several investigations have demonstrated structural and functional changes, by using magnetic resonance imaging, eletroencephalography or magnetoencephalography, in brains of musicians as a result of many years of musical practice.

In examining the morphometry of the corpus callosum, Schlaug et al. (1995) found that its anterior portion was significantly larger in musicians compared to non-musicians. Moreover, a comparison between subgroups revealed that this region was larger in those who began musical training earlier. The authors pointed out that the observed anatomical difference has to be taken in the context of a need for increased interhemispheric communication underlying complex bimanual motor sequences in musicians.

Amunts et al. (1997) evaluated the size of right and left motor cortices in right-handed musicians and non-musicians. As expected, a leftward asymmetry was observed in both groups, but musicians showed a smaller degree of asymmetry, since they presented a larger right motor cortex. The investigators also found a negative correlation between the size of motor cortex of both hemispheres and the age of commencement of musical practice. These authors interpreted the structural differences in motor cortex as structural compliance in response to intense and early hand skill training.

In a retrospective study Schlaug et al. (1998) measured the volume of cerebellum in musicians and non-musicians, showing higher average relative cerebellar volume in male musicians when compared to male non-musicians, a finding interpreted as evidence for microstructural adaptations in the cerebellum in response to early commencement and continual practice of complicated bimanual finger sequences.

Gaser and Schlaug (2003) found increased gray matter volume in motor, auditory and visuospatial cerebral areas in musicians, using voxel-based morphometry analysis, a method enabling structural differences to be identified across the whole brain space. According to the authors, the results can be viewed as evidence of structural adaptations in response to long-term skill acquisition and the repetitive rehearsal of these skills, a notion supported by the strong association found by the authors between structural differences and practice intensity.

Bengtsson et al. (2005) using diffusion tensor imaging, investigated effects of extensive piano practice on white matter in childhood, adolescence and adulthood. They found more structured right posterior internal capsule, which carries corticospinal tracts, in pianists compared to non-musicians, and also found positive correlations between practicing and fiber tract organization in different regions for each age period. The authors noted that training can induce white matter plasticity if this occurs during periods when fiber tracts involved are still undergoing maturation.

Han et al. (2009) examined both gray matter and white matter in pianists and non-musicians and demonstrated, in the first group, higher gray matter density in the left primary sensory-motor cortex and right cerebellum as well as higher white matter integrity in the right posterior internal capsule. According to the researchers, these results indicate that long-term piano practicing may lead to gray matter and white matter adaptation in movement-related regions, which may impact number of synapses, volume of glia or increased myelination and diameter of axons.

MUSIC AND FUNCTIONAL CHANGES IN THE BRAIN

Music listening activates a wide-spread bilateral network of brain regions. Recent brain imaging studies have shown that neural activity associated with music listening extends well beyond the auditory cortex involving a wide-spread bilateral network of frontal, temporal, parietal and subcortical areas related to attention, semantic and music-syntactic processing, memory and motor functions (Bhattacharya et al., 2001; Janata et al., 2002; Popescu et al., 2004), as well as limbic and paralimbic regions related to emotional processing (Blood et al., 1999; Blood and Zatorre, 2001; Menon and Levitin,

2005).Other studies have shown differences in functional cerebral characteristics between musicians and non-musicians.

Elbert et al. (1995) investigated the somatosensory cortical representation of the fingers D1 (thumb) and D5 (little finger) of string players – violinists, cellists, guitarists – and non-musicians. After excitation of left-hand fingers D1 and D5, the strength of cortical activation was higher in musicians than in control subjects. This effect was particularly pronounced for finger D5, while there was no difference in cortical representations following right-hand stimulation. The authors stated that cerebral representation is enhanced to the fingers of the hand that is most intensively used in string players, and that the more a given finger is stimulated, the larger the increase in cortical response.

Pantev et al. (1998) investigated auditory cortical representation in musicians and non-musicians. After acoustic stimulation, consisting of a pseudorandom sequence of four piano tones and four pure tones matched in frequency and loudness, the strength of cortical activation for piano tones was 25% greater in musicians. There was no difference between the groups in relation to cortical representation of pure tones. The authors stressed that pure tones, unlike piano tones, are not part of our natural acoustic environment and also are not commonly encountered in musical training and practice, which may account for the observed results. Both Elbert et al. (1995) and Pantev et al. (1998) revealed a greater degree of cortical representation in musicians who began musical training early.

Herdener et al. (2010) investigated plastic capabilities of the hippocampus by evaluating brain responses induced by temporal novelty in musicians and non-musicians, since the hippocampus structure, in addition to its key role for memory and spatial navigation, has been suggested to be crucially involved in various forms of novelty detection. They observed enhanced neural responses to temporal novelty in the anterior left hippocampus in professional musicians (cross-sectional study) and in music students after one year of intensive aural skills training (longitudinal study). The authors also found a correlation between hippocampal sensitivity to temporal novelty and musical abilities. They assumed that the observed changes in hippocampal activity in musicians represented a functional correlate of a tuning of aural skills, related to time interval perception, during the course of their studies.

Musacchia et al. (2007) showed that changes in functional organization also extend to subcortical sensory structures, demonstrating that musicians, compared to controls, had earlier and larger auditory and audiovisual brainstem responses to speech and music stimuli. The data also showed a positive correlation between years of musical practice and strength of brainstem response to speech stimulus, suggesting that musicians acquire an enhanced representation of pitch through musical training. All of the previously addressed studies suggested the existence of neuroplastic processes as an effect of musical training.

In a longitudinal study, Altenmüller (2000) demonstrated that cortical activation during music processing reflects the auditory "learning biography", the personal experiences accumulated over time. The author proposed a model to represent the relationship between auditory information and neural networks involved in music processing. According to his model, complexity of neural networks enhances with the complexity of auditory information. More interestingly, musical training can add mental representations of music, which may involve different cerebral substrates. These representations can be auditory, sensory-motor, symbolic, visual, among others. Therefore, according to the author, for the same level of auditory information complexity, professional musicians presumably use larger and more complex neural networks compared to non-musicians.

Music and Language

The domains of music and language share many features, the most direct being that both exploit changes in pitch patterns to convey information. Music uses pitch contours and intervals to communicate melodies and tone centers. Pitch patterns in speech convey prosodic information; listeners use prosodic cues to identify indexical information, i.e., information about the speaker's intention as well as emotion and other social factors. Further, in tonal languages, changes in pitch are used lexically; that is, in differentiating between words. (e.g., In Mandarin: /ma/ (high level) means 'mother', /ma/ (high rising) means 'hemp' , /ma/ (low falling rising) means 'horse', ma (high falling) means 'scold').

A significant body of research has focused on the extent to which musical experience provides benefits in language abilities; the results unambiguously suggest that musicians show enhanced processing of prosodic and linguistic pitch. Musicians show an enhanced ability to detect subtle incongruity in prosodic pitch as well as consistent neural differences relative to non-musicians (Besson, Schon, Moreno, Santos & Magne, 2007; Magne, Schon, & Besson, 2006). Musicians showed superior cortical representation of linguistic pitch in a non-native language relative to non-musicians (Krishnan et al., 2005; Chandrasekaran et al. 2009). The ability to track non-native pitch contours correlated positively with number of years of musical training, suggesting that it was musical experience that improved lower level representation of non-native pitch. Musicians showed a superior propensity to use pitch in lexical contexts during a language learning task, relative to non-musicians (Wong & Perrachione, 2007).

Are musicians better language learners? Research has found that children who study music before the age of seven develop bigger vocabularies, a better sense of grammar and a higher verbal IQ. According to the studies, just one hour a week of learning music is enough for the full brain benefits to take place – including an all-round boost in language skills and a significant increase in IQ. Children who received instrumental training not only showed enhanced processing of skills related to music, but also showed enhanced vocabulary relative to untrained controls (Forgeard, Winner et al., 2008).

Children who learn music from a young age find it easier to learn languages even in adulthood, research has found. As music training boosts all the language-related networks in the brain, we can expect it to be beneficial in the acquisition of languages, and this is what the studies have found. These advantages benefit both the development of their mother tongue and the learning of foreign languages. During these first few years, the brain is at its sensitive development phase, with 95% of the brain's growth occurring then. Music training started during this period also boosts the brain's ability to process subtle differences between sounds and assist in the pronunciation of languages – and this gift lasts for life, as it has been found that adults who had musical training in childhood still retain this ability to learn foreign languages quicker and more efficiently than adults who did not have early childhood music training.

Finnish children are commonly musically trained from a young age (up until the age of seven) with the playful Musiikkileikkikoulu method, but they only start school at age seven and start language learning at nine or older. Despite this "late exposure" to everything excluding music skills, they commonly end up speaking three to five foreign languages. Any English-speaking person who has ever visited Finland can attest to the fact that nearly every Finnish person speaks English without any problems.

Humans first started creating music 500,000 years ago, yet speech and language was only developed 200,000 years ago. Evolutionary evidence, as interpreted by leading researchers such as Robin Dunbar from Oxford University, indicates that speech as a form of communication has evolved from our original development and use of music. This explains why our music and language neural networks have significant overlap, and why children who learn music become better at learning the grammar, vocabulary and pronunciation of any language.

Research has indicated that musical training may serve as a useful remediation strategy for children with language impairments (Overy et al., 2003; Besson et al., 2007; Jentschke et al., 2008; Jentschke & Kolesch., 2009). Cunningham et al., 2001, indicted that clinical population known to have problems with language based learning disabilities (e.g., poor readers), may also get benefit from musical training.

Music and Academics

There is now new evidence that an early education in music can help with concentration, coordination and now even maths and science. Researchers also have found a significant relationship between music instruction and positive performances in such areas as: reading comprehension, spelling, mathematics, listening skills, primary mental abilities and motor skills (Borgese, 2010).

In typically developing children with normal reading ability, musical discrimination skills significantly predicted phonological and reading skills (Forgeard, Schlaug et al., 2008). Hallam (2002) showed that elementary school students who listened to mood-calming music while completing mathematical problems were able to complete more problems and solve a higher percentage of them correctly than the group who listened to no music at all.

The effect demonstrated that there may be an important relationship between certain types of music (e.g. classical) and learning (Jackson & Tluaka, 2004). One study involving college students showed a correlation between how awake they felt and their preference for music or silence. Results indicated a positive effect while listening to Mozart (Jones, West, & Estell, 2006). This effect has become known as the Mozart Effect, which proposes that listening to Mozart can increase spatial abilities. The proposed increase in the construction of alpha waves may result in positive learning ability.

Music and Emotion

The subject of music and emotions is one of the most heavily researched in the field of music psychology. There is the debate surrounding whether music really stimulates actual emotions, but some researchers argue that it does (at least in part) because musical emotions have all the signs of 'real' emotions including bodily reactions, facial expressions, and action responses (such as dancing, singing, and crying). Another interesting source of emotion is referred to as 'contagion': The idea that once an emotion is triggered, we experience the physiological manifestations of that state – so, for example, we might smile. That smile then feeds back into the system and reinforces the happy emotion that we feel.

Perception of emotion in speech and music relies on shared acoustic and neural mechanisms (Nair et al., 2002), suggesting that extensive experience in one domain may lend perceptual benefits to the other. Relative to non-musicians, musicians showed superior encoding of the most acoustically complex portion of the emotional stimuli, consistent with behavioral studies demonstrating enhanced emotional perception in musicians (Thompson, Schellenberg & Husain, 2004).

Examining the subcortical encoding of a complex for emotionally salient stimulus (a Child's cry) as a function of music experience, a recent study demonstrated increased neural efficiency in musicians (Strait et al., 2009; Strait, Kraus, Skoe & Ashley, 2009). In this study they aimed to provide a biological basis for musician's enhanced perception of emotion in speech by investigating the contribution of subcortical mechanisms to the processing of vocally communicated emotional states. 30 musicians were included in the studies, they were classified into 2 groups based on 2 criteria: musicians by onset age (MusAge) and musicians by years (Mus Yrs). MusAge subjects had begun musical training at or before age of 7 years, whereas Mus Yrs subjects had received more than 10 years of consistent musical experience. Integrity of auditory brainstem was assessed using auditory brainstem responses with both click and speech (/da/). The authors suggested that musical experience has more pervasive domain-general effects on the auditory system than previously documented, resulting in fine neural timing to acoustic features important for vocal communication. The results thus provide evidence for initial biologic involvement of subcortical mechanisms in the auditory processing of communicated states of emotion.

Music and Cognition

Listening to music is a complex process for the brain, since it triggers a sequel of cognitive and emotional components with distinct neural substrates (Peretz and Zatorre, 2005). Further, playing a musical instrument is a complex and motivating activity that comprises the coordination of multiple sensory modalities (auditory, visual, and somatosensory) and motor system in a unique way. Several investigations have shown that structural and functional cerebral neuroplastic processes emerge as a result of long-term musical training, which in turn may produce cognitive differences between musicians and non-musicians. Recent cognitive and neuropsychological studies suggest that music learning enhances a variety of cognitive functions, such as attention, learning, communication and memory, both in healthy participants (Wallace, 1994; Thompson et al., 2001; Thompson et al., 2005; Schellenberg et al., 2007) and in patients, such as dyslexia (Overy, 2003), autism (Gold et al., 2006), schizophrenia (Talwar et al., 2006), multiple sclerosis (Thaut et al., 2005), coronary artery disease (Emery et al., 2003) and dementia (Brotons and Koger, 2000; Foster and Valentine, 2001; Van de Winckel et al., 2004).

Does music make you smarter? Music listening and music lessons have been claimed to confer intellectual advantages. Schellenberg (2004) found that children who received 36-weeks musical lessons (standard keyboard or Kodálay) showed a small but significant increase in IQ compared to children who took drama lessons or no lessons at all.

Music training also has been shown to improve working memory (Forgeard, Winner et al., 2008; Jakobson, Lewycky, Kilgour, & Stoesz, 2008; Parbery-Clark, Skoe, Lam et al., 2009; and executive function abilities (Bialystok & DePape, 2009. Numbers of studies have evidenced a musician enhancement for auditory, visual and verbal memory. In a hallmark study, Chan and colleagues showed that participants with music training exhibited superior verbal memory relative to non-musicians, as indicated by greater number of words recalled in a list learning task (Chan et al., 1998).

Jakobson et al. (2008) in a part of their work, studied visual memory in pianists and non-musicians by using the Rey Visual Design Learning Test (RVDLT). In this test, participants try to learn and remember a sequentially presented set of 15 line drawings of simple geometric figures, each containing two elements (e.g., a circle and a line). Recall, registered by asking subjects to draw all figures they can remember, is tested after each of five learning trials and following a delay period. A test of delayed recognition is also administered. The results suggested superior visual memory in musicians, since they outperformed non-musicians on the fourth and fifth learning trials and on the delayed recall and delayed recognition tasks. After controlling statistically for general intelligence, the group difference on the delayed recall tasks persisted.

Rodrigues et al. (2007) compared the performance of musicians, members of a symphony orchestra and a symphony band, and non-musicians, in tasks involving visual attention ability. The main neuropsychological test used in the study was the Multiple Choice Reaction Time (MCRT) test, which consisted of specific motor responses to various luminous stimuli. In order to evaluate divided visual attention ability, the MCRT test was applied twice, the second time concomitantly with other continuously and randomly changing visual stimuli presented in video form. Subjects were asked to respond verbally to each change. Musicians showed a higher percentage of correct responses to MCRT when the test was applied alone. Although no significant difference of accuracy was observed between the groups when MCRT was applied together with changing visual stimuli, musicians showed shorter reaction times for verbal responses to stimuli changes. The authors pointed out that this result may suggest augmented divided visual attention ability in musicians compared to non-musicians, which they ascribed to ensemble musical practice. The professional routine of musicians is characterized by constant demands of divided visual attention, since it requires dealing with several kinds of visual stimuli simultaneously such as the music score, conductor's gestural instructions and body movements of other musicians, while playing the instrument.

Ho et al. (2003) used a hybrid cross-sectional and longitudinal design and assessed verbal memory ability in children with and without musical training using the Hong Kong List Learning Test (HKLLT). In the cross-sectional part of the study, the results showed that children with musical training demonstrated better verbal memory than those without such training. The longitudinal part of the study consisted of a one-year follow-up of the children, forming three groups: those with experience who continued lessons, those who discontinued their lessons, and those who were new to musical training. The researchers verified that children who had begun or continued musical training demonstrated significant verbal memory improvement and those who had discontinued the training showed no improvement at all. They associated their results with the neuroimaging findings of Schlaug et al. (1995) that revealed an enlargement in the left planum temporale in musicians.

Music and Health

In general healthcare settings, the relaxing qualities of listening to music have been used to complement pain management techniques, to raise patients' mood, promote movement for physical rehabilitation, to relieve sleeplessness, calm nerves and reduce muscle tension and many more advantages. Studies have found that post-operative patients can ease their pain and reduce their dependence on pain-killers by listening to one of the famous Ragas of Carnatic music, Anandha Bhairavi. Music has a well-documented effect on alleviating anxiety, depression and pain in patients with a somatic illness (Cassileth et al., 2003; Cepeda et al., 2006; Siedliecki and Good, 2006).

Australian and international research has shown that listening to music reduces anxiety while waiting for invasive hospital procedures, and patients undergoing chemotherapy while listening to music showed an improvement in levels of anxiety, fear fatigue and blood pressure. Other research indicates that listening to music reduces anxiety and depression felt by cardiac bypass surgery patients when recovering from surgery. Music is often used as a tool during labour and childbirth, for focus or relaxation. Music therapy is a creative way for people with HIV/AIDS to discover, explore, and work through personal issues related to their condition.

Verghese et al. (2003), in a follow-up study of elderly people, observed that those individuals who played a musical instrument were less likely to suffer dementia than participants involved in other type of leisure activities like reading, writing, or doing crossword puzzles. Active and passive musical activities in elderly people have been found to improve mood and reduce depression symptoms (Chan et al., 2010; Erkkila et al., 2011). Piano lessons decreased depression, induced positive mood states, and improved the psychological and physical quality of life of the elderly. Our results suggest that playing piano and learning to read music can be a useful intervention in older adults to promote cognitive reserve and improve subjective well-being.

Elements of music have previously been used as a part of physiotherapy (Thaut et al., 1997) and speech therapy (Belin et al., 1996) to enhance the recovery of motor and speech functions such as dyslexia (Overy, 2003), autism (Gold et al., 2006), schizophrenia (Talwar et al., 2006), multiple sclerosis (Thaut et al., 2005), coronary artery disease (Emery et al., 2003) and dementia (Brotons and Koger, 2000; Foster and Valentine, 2001; Van de Winckel et al., 2004). The benefits of musical training have been shown in a stroke patient evidenced by a change in the reorganization of the sensorimotor cortex and an improvement in his movement quality after receiving music-supported therapy (Rojo et al., 2011; Rodriguez-Fornells et al., 2012).

Music therapy is an important tool used to help children with learning disabilities. Music therapy can help to capture the attention of children with learning disabilities in a way that many other mental and physical activities cannot. It has been used as an effective way to teach children with learning disabilities in a way that is easy and clear for them to understand. There are various musical activities that can be used to help these children. It is often the case that children who suffer from stuttering and speech disorders are able to sing fluently although they have great difficulty when speaking. Many children and adults who stutter find it easier to talk if they use a sing-song style of talking. The rhythm and pace of using a sing-song technique makes it easier for speech to flow.

Music therapy can be extremely effective when working with autistic children, as they can be extremely sensitive to music. Some autistic children are able to sing, despite not being able to speak. In these cases, music therapy uses simple songs with repetitive phrases to help to develop a child's language and speech. Some autistic children have a remarkable talent for music, for example, some have perfect pitch while others have an outstanding ability to play a musical instrument. A 'rock music backdrop' has been successfully used with Attention Deficit Hyperactive Disorder (ADHD) children to help them focus.

On a concluding note, the power of music can be stated with the quote given by Berthold Auerbach as "Music washes away from the soul the dust of everyday life."

Music and Dance

Dancers perform to music and there involves passive music learning. Thus, it is interesting to understand if perception of music aspects in dancers is similar to musicians. Rhythm is one aspect of music on which dancers also depend on for their performance. The Bharatanatyam dance form and Carnatic music share similar musical compositions and are exposed to the same rhythms and they are trained similarly. But while performing and practicing, rhythm is maintained in different manner by the two groups. The Carnatic musicians maintain rhythm by tapping with the palm and Bharatanatyam dancers maintain practice by stamping it on their feet.

Arya and Rajalakshmi (2012) compared the rhythm perception skills of Carnatic musicians and Bharathanatyam dancers, to find out whether perception of beats influences identification and synchrony with the rhythm of a musical composition The result implied that the rhythm perception abilities in dancers and musicians are comparable. Subjects in both the groups could identify the rhythm; synchronize with the musical composition according to its phase and tempo to the same extent.

Early musical training helps develop brain areas involved in language and reasoning. It is thought that brain development continues for many years after birth. Recent studies have clearly indicated that musical training physically develops the part of the left side of the brain known to be involved with processing language, and can actually wire the brain's circuits in specific ways. There is also a causal link between music and spatial intelligence which is critical to the sort of thinking necessary for everything from solving advanced mathematics that will be needed for the day. Through music study, students learn the value of sustained effort to achieve excellence and the concrete rewards of hard work.

Music study enhances teamwork skills and discipline. Music provides children with a means of self-expression. Music study develops skills that are necessary in the workplace. It focuses on "doing," as opposed to observing, and teaches students how to perform literally anywhere in the world. Music performance teaches young people to conquer fear and to take risks.

REFERENCES

Akbaraly, T. N., Portet, F., Fustinoni, S., Dartigues, J.-F., Artero, S., Rouaud, O., et al. (2009). Leisure activities and the risk of dementia in the elderly: results from the Three-City Study. *Neurology, 73,* 854–861.

Amunts, K., Schlaug, G., & Jancke, L. et al. (1997). Motor cortex and hand motor skills: structural compliance in the human brain. *Human Brain Mapping, 5,* 206-215.

Anvari, S., Trainor, L., Woodside, J., & Levy, B. (2002).Relations among musical skills, phonological processing and early reading ability in preschool children. *Journal of Experimental Child Psychology, 83,*111-130.

Arya, C., & Rajalakshmi, K., (2012). *Comparison of Rhythm perception in Dancers and Musicians.* Published Master's Dissertation, University of Mysore, Mysore.

Baeck, E. (2002). The neural networks of music. *European Journal of Neurology, 9,* 449-456.

Bengtsson, S., Nagy, Z., Skare, S., Forsman, L., Forssberg, H., & Ullén, F. (2005). Extensive piano practicing has regionally specific effects on white matter development. Nature Neuroscience, 8, 1148-1150.

Besson, M., Chobert, J., & Marie, C. (2011). Transfer of training between music and speech: common processing, attention, and memory. *Frontiers in Psychology, 2*, 94.

Besson, M., Faita, F., & Requin, J. (1994). Brain waves associated with musical incongruities differ for musicians and non-musicians. *Neuroscience Letter, 168,* 101-105.

Bever, T.G, & Chiarello, R.J. (1974).Cerebral dominance in musicians and non-musicians. *Science, 185,* 537-539.

Bilhartz, T., Bruhn, R., & Olson, J. (1999).The effect of early music training on child cognitive development. *Journal of Applied Developmental Psychology, 20,* 615-636.

Bugos, J. A., Perlstein, W. M., McCrae, C. S., Brophy, T. S., & Bedenbaugh, P. H. (2007). Individualized piano instruction enhances executive functioning and working memory in older adults. *Aging Mental Health, 11,* 464–471.

Chan, A. S., Ho, Y. C., & Cheung, M.C. (1998). Music training improves verbal memory. *Nature, 396,*128.

Chan, M. F., Chan, E. A., & Mok, E. (2010). Effects of music on depression and sleep quality in elderly people: a randomised controlled trial. *Complementary Therapies in Medicine, 18,* 150–159.

Chobert, J., François, C., Velay, J.-L., & Besson, M. (2012). Twelve months of active musical training in 8- to 10-year-old children enhances the preattentive processing of syllabic duration and voice onset time. *Cerebral Cortex, 23,* 3874–3887.

Costa-Giomi, E. (1999).The effects of three years of piano instruction on children's cognitive development. *Journal of Research in Music Education, 47,* 198-212.

Creech, A., Hallam, S., McQueen, H., & Varvarigou, M. (2013). The power of music in the lives of older adults. *Research Studies in Music Education. 35,* 87–102.

Elbert, T., Pantev, C., Wiendbruch, C., Rockstroh, B., & Taub, B. (1995).Increased cortical representation of the fingers of the left hand in string players. *Science,* 270, 305-307.

Erkkila, J., Punkanen, M., Fachner, J., Ala-Ruona, E., Pontio, I., Tervaniemi, M., et al. (2011). Individual music therapy for depression: randomised controlled trial. *British Journal of Psychiatry, 199,* 132–139.

Forgeard, M., Winner, E., Norton, A., & Schlaug, G. (2008). Practicing a musical instrument in childhood is associated with enhanced verbal ability and nonverbal reasoning. *PLoS One, 3,* 1-8.

François, C., Chobert, J., Besson, M., & Schön, D. (2013).Music training for the development of speech segmentation. *Cerebral Cortex, 23,* 2038–2043.

Franklin, M., Moore, K., Yip, C.Y., Jonides, J., Rattray, K., & Moher, J. (2008).The effects of musical training on verbal memory. *Psychology of Music, 36,* 53-365.

Gaser, C., & Schlaug, G. (2003). Brain structures differ between musicians and non-musicians. *Journal of Neuroscience, 23,* 9240–9245.

Graziano, A., Peterson, M., & Shaw, G. (1999).Enhanced learning of proportional math through music training and spatial-temporal reasoning. Neurology Research, 21, 139-152.

Gromko, J. (2005). The effect of music instruction on phonemic awareness in beginning readers. *Journal of Research in Music Education, 53,* 199-209.

Han, Y., Yang, H., Lv, Y.T. et al. (2009). Gray matter density and white matter integrity in pianists' brain: a combined structural and diffusion tensor MRI study. *Neuroscience Letter, 459,* 3-6.

Hanna-Pladdy, B., & MacKay, A. (2011).The relation between instrumental musical activity and cognitive aging. *Neuropsychology, 25,* 378–386.

Herdener, M., Esposito, F., di Salle, F., et al. (2010). Musical training induces functional plasticity in human hippocampus. *Journal of Neuroscience, 30,* 1377-1384.

Herholz, S. C., & Zatorre, R. J. (2012). Musical training as a framework for brain plasticity: behavior, function, and structure. *Neuron, 76,* 486–502.

Hetland, L. (2000). Learning to make music enhances spatial reasoning. *Journal of Aesthetic Education, 34,* 179-238.

Hirshkowitz, M., Earle, J., & Paley, B. (1987). EEG alpha asymmetry in musicians and non-musicians :a study of hemispheric specialization. *Neuropsychologia, 16,* 125-128.

Ho, Y., Cheung, M., & Chan, A. (2003). Music training improves verbal but not visual memory: cross sectional and longitudinal explorations in children. *Neuropsychology, 17,* 439-450.

Hyde, K. L., Lerch, J., Norton, A., Forgeard, M., Winner, E., Evans, A. C., et al. (2009). Musical training shapes structural brain development. *Journal of Neuroscience, 29,* 3019–3025.

Hyde, K., Lerch, J., Norton, A, et al. (2009). Musical training shapes structural brain development. *Journal of Neuroscience, 29,* 3019-3025.

Jakobson, L., Lewycky, S., Kilgour, A., & Stoesz, B. (2008). Memory for verbal and visual material in highly trained musicians. *Music Perception, 26,* 41-55.

Koelsch, S. (2014) Brain correlates of music-evoked emotions. *Nature Reviews Neuroscience*, 15, 170-180.

Koelsch, S., Schroger, E., & Tervaniemi, M. (1999).Superior pre-attentive auditory processing in musicians. *NeuroReport, 10,* 1309-1313.

Kraus, N., & Chandrasekaran, B. (2010).Music training for the development of auditory skills. *Nature Review Neuroscience, 11,* 599–605.

Lappe, C., Herholz, S. C., Trainor, L. J., & Pantev, C. (2008). Cortical plasticity induced by short-term unimodal and multimodal musical training. *Journal of Neuroscience, 28,* 9632–9639.

Lappe, C., Trainor, L. J., Herholz, S. C., & Pantev, C. (2011). Cortical plasticity induced by short-term multimodal musical rhythm training. *PloS One, 6* .

Lee, Y. Y., Chan, M. F., & Mok, E. (2010).Effectiveness of music intervention on the quality of life of older people. *Journal of Advanced Nursing, 66,* 2677–2687.

Moreno, S., Marques, C., Santos, A., Santos, M., Castro, S. L., & Besson, M. (2009). Musical training influences linguistic abilities in 8-year-old children: more evidence for brain plasticity. *Cerebral Cortex, 19,* 712–723.

Münte, T., Altenmuller, E., & Jancke, L. (2002).The musician's brain as a model of neuroplasticity. *Nature Reviews Neuroscience, 3,* 473-478.

Musacchia, G., Sams, M., Skoe, E., & Kraus, N. (2007). Musicians have enhanced subcortical auditory and audiovisual processing of speech and music. *Proceedings of National Academy of Sciences, 104,* 15894-15898.

Ohnishi, T., Matsuda, H., Asada, T., et al. (2001).Functional anatomy of musical perception in musicians. *Cerebral Cortex, 11,* 754-760.

Pantev, C., Oostenveld, R., Engelien, A., Ross, B., Roberts, L., & Hoke, M. (1998). Increased auditory cortical representation in musicians. *Nature, 392,* 811-813.

Peretz, I. (2006). The nature of music from a biological perspective. *Cognition, 100,* 1-32.

Piro, J., & Ortiz, C. (2009).The effect of piano lessons on the vocabulary and verbal sequencing skills of primary grade students. *Psychology of Music, 37,* 1-23.

Platel, H., Price, C., Baron, J. C., et al. (1997). The structural components of music perception: a functional anatomic study. *Brain, 120,* 229-243.

Rauscher, F., & Zupan, M. (2000). Classroom keyboard instruction improves kindergarten children's spatial-temporal performance: a field experiment. *Early Childhood Research Quarterly, 15,* 215-228.

Rodrigues, A., Guerra, L., & Loureiro, M. (2007). Visual attention in musicians and non-musicians: a comparative study. *Proceedings of the 3rd International Conference on Inter- disciplinary Musicology,* Tallinn, Estonia.

Rodriguez-Fornells, A., Rojo, N., Amengual, J. L., Ripollés, P., Altenmüller, E., & Munte, T. F. (2012). The involvement of audio-motor coupling in the music-supported therapy applied to stroke patients. *Annals of New York Academy of Sciences, 1252,* 282–293.

Rojo, N., Amengual, J., Juncadella, M., Rubio, F., Camara, E., Marco-Pallares, J., et al. (2011). Music-supported therapy induces plasticity in the sensorimotor cortex in chronic stroke: a single- case study using multimodal imaging (fMRI-TMS). *Brain Injury, 25,* 787–793.

Rüsseler, J., Altenmüller, E., Nager, W., Kohlmetz, C., & Münte, T. (2001). Event-related brain potentials to sound omissions differ in musicians and non-musicians. *Neuroscience Letter, 308,* 33-36.

Schellenberg, E.G. (2004). Music lessons enhance IQ. *Psychology of Science, 15,* 511–514.

Schellenberg, E.G., & Peretz, I. (2007). Music, language and cognition: unresolved issues. *Trends in Cognitive Sciences, 12,* 45-46.

Schlaug, G., Jancke, L., Huang, Y., & Steinmetz, H. (1995).In vivo evidence of structural brain asymmetry in musicians. *Science, 267,* 699-701.

Schlaug, G. (1995). Increased corpus callosum size in musicians. Neuropsychologia 33, 1047–1055.

Schlaug, G., Jäncke, L., Huang, Y., Staiger, J., & Steinmetz, H. (1995).Increased corpus callosum size in musicians. *Neuropsychologia, 33,* 1047-1055.

Schlaug, G., Lee, L., Thangaraj, V., et al. (1998). Macrostructural adaptation of the cerebellum in musicians. *Social Neuroscience, 24,* 842-847.

Schmithorst, V. J., & Wilke, M. (2002). Differences in white matter architecture between musicians and non-musicians: a diffusion tensor imaging study. *Neuroscience Letter, 321,* 57–60.

Schmithorst, V.J., & Holland, S.K. (2003). The effect of musical training on music processing: a functional magnetic resonance imaging study in humans. *Neuroscience Letter,* 348, 65-68.

Sluming, V., Barrick, T., Howard, M., Cezayirli, E., Mayes, A., & Roberts, N. (2002). Voxel-based morphometry reveals increased gray matter density in Broca's area in male symphony orchestra musicians. *NeuroImage, 17,* 1613-1622.

Solé, C., Mercadal-Brotons, M., Gallego, S., & Riera, M. (2010).Contributions of music to aging adults' quality of life. *Journal of Music Therapy, 47,* 264–281.

Standley, J., & Hughes, J. (1997).Evaluation of an early intervention music curriculum for enhancing prereading/writing skills. *Music Therapeutic Perspective, 15,* 79-85.

Stewart, L., Henson, R., Kampe, K., Walsh, V., Turner, R., and Frith, U. (2003). Brain changes after learning to read and play music. *Neuroimage, 20,* 71–83.

Travis, F., Harung, H. S., & Lagrosen, Y. (2011). Moral development, executive functioning, peak experiences and brain patterns in professional and amateur classical musicians: interpreted in light of a Unified Theory of Performance. *Consciousness and Cognition, 20,* 1256–1264.

Vaughn, K. (2000). Music and mathematics: modest support for the oft-claimed relationship. *Journal of Aesthetic Education, 34,* 149-166.

Verghese, J., Lipton, R. B., Katz, M. J., Hall, C. B., Derby, C. A., Kuslansky, G., et al. (2003). Leisure activities and the risk of dementia in the elderly. *New England Journal of Medicine, 348,* 2508–2516.

Wan, C. Y., & Schlaug, G. (2010). Music making as a tool for promoting brain plasticity across the life span. *Neuroscientist, 16,* 566–577.

Chapter 5

BENEFITS OF MUSIC ON AUDITORY PROCESSING

From the cochlea to the auditory cortex, sound is encoded at multiple locations along the ascending auditory pathway, eventually leading to conscious perception (Kraus 2007). Speech is a stream of acoustic elements produced at an astounding average rate of three to six syllables per second (Laver, 1994). The ability to decode these elements in a meaningful manner is a complex task that involves multiple stages of neural processing.

Neural plasticity is a term used to describe alterations in the physiological and anatomical properties of neurons in the brain in association with auditory stimulation and deprivation. Depending on the experience, mechanism of plasticity can involve synaptic changes that occur rapidly or slowly over a period of time (Tremblay & Kraus, 2001). Everyday learning and training involves of continuous improvement of our abilities in the sensory, cognitive & behavioural levels (Menning, Roberts & Pantev, 2000). Peripheral and central structures along the auditory pathway contribute to speech processing and learning.

Music is a complex auditory task and musicians spend years fine-tuning their skills. It is no wonder that previous research has documented neuroplasticity to musical sounds as a function of experience (Fujioka, Trainor, Ross, Kakigi, Pantev, 2005; Koelsch, Schroger, & Tervaniemi, 1999; Musacchia, Sams, Skoe, & Kraus, 2007; Pantev et al., 1998; Pantev, Roberts, Schulz, Engelien, & Ross, 2001; Tervaniemi, Rytkonen, Schroger, Ilmoniemi, & Naatanen, 2001). Many studies have reported that musicians have better auditory perception skills when compared to non-musicians. There are many studies in literature which have documented that musical training improves basic auditory perceptual skills resulting in enhanced behavioral (Jeon & Fricke, 1997; Koelsch et al., 1999; Oxenham et al., 2003; Tervaniemi et al., 2005; Micheyl et al., 2006; Rammsayer & Altenmuller, 2006) and neurophysiological responses (Brattico et al., 2001; Pantev et al., 2001; Schneider et al., 2002; Shahin et al., 2003 & 2007; Tervaniemi et al., 2005; Kuriki et al., 2006; Kraus et al., 2009).

BEHAVIOURAL STUDIES AS EVIDENCE FOR ENHANCEMENT IN AUDITORY PERCEPTION

Speech in Noise perception and Temporal Processing Abilities

Speech perception in noise (SIN) is a complex task requiring the segregation of the target signal from the competing background noise. This task is further complicated by the degradation of the acoustic signal, with the noise particularly disrupting the perception of the fast spectro-temporal changes (Brandt & Rosen, 1980). Whereas children with language-based learning disabilities (Bradlow et al., 2003; Ziegler et al., 2005) and hearing impaired adults (Gordon- Salant & Fitzgibbons, 2005) are especially susceptible to the negative effects of background noise, musicians are less affected and demonstrate better performance for SIN when compared to non-musicians. As a consequence of the musician's extensive experience with auditory stream analysis within the context of music, more honed auditory perceptual skills, musicians seem well equipped to cope with the demands of adverse listening situations such as Speech in Noise.

Musicians, as a consequence of training that requires consistent practice, online manipulation, and monitoring of their instrument, are experts in extracting relevant signals from the complex sound scape (e.g., the sound of their own instrument in an orchestra). Literature shows that the effect of musical experience is transferred on the skills that sub serve successful perception of speech in noise.

Rammsayer and Altenmuller (2006) studied 36 academically trained musicians and 36 non-musicians. Seven different auditory temporal tasks were performed in the study. Auditory fusion, rhythm perception, and three temporal discrimination abilities were found to be superior for musicians when compared to non-musicians. The authors reported that temporal information processing is more accurate in musicians than in non-musicians.

Ishll et al. (2006) showed that music has a positive influence on the development of the planum temporale, because according to their study, subjects who were exposed to musical training (singing) over four years compared to amateur musicians without professional guidance, performed better on temporal resolution through the test Random Gap Detection Threshold (RGDT).

Study by Parbery-Clark, Skoe, Lam et al.(2009) found a distinct speech in noise advantage for musicians, as measured by standardized tests of hearing in noise (HINT, Hearing in- noise test; QuickSIN). Across all participants, the number of years of consistent practice with a musical instrument correlated strongly with performance on QuickSIN, auditory working memory and frequency discrimination. These correlations strongly suggest that practice fine tunes cognitive and sensory abilities, leading to an overall advantage for speech perception in noise in musicians. The results from the study suggest that musical experience enhances the ability to hear speech in challenging listening environments. SIN performance is a complex task requiring perceptual cue detection, stream segregation, and working memory. Musicians performed better than non-musicians in conditions where the target and the background noise were presented from the same source, meaning parsing was more reliant on the acoustic cues present in the stream.

Mohamdkhani, Nilforoushkhoshk, Mohammadi, Faghihzadeh and Sepehrenejhad (2010) conducted a study on 24 musicians and 24 normal hearing non-musician controls. GIN (Gap in Noise) test results indicated that there was a significant difference between approximate threshold and percent of corrected answers between musicians and non-musician group. They concluded that musicians had rapid auditory temporal processing ability as compared to the non-musicians group as the musician group showed lower approximate threshold and the more corrected answers in GIN test. The authors attribute this to the effect of musical training on central auditory processing.

Guclu, Sevinc and Canbeyli (2011), studied duration discrimination thresholds, musicians were found to have on average 10% better thresholds than non-musicians.

Thomas and Rajalakshmi (2011) compared temporal resolution abilities and speech perception in noise in 20 trained carnatic vocal musicians across their years of experience (5 years to 16 years). Temporal resolution abilities were measured using the temporal modulation transfer function (TMTF) and Gap detection threshold test. TMTF was measured for 6 different modulation frequencies, for both the ears separately. The results from the present study showed that the temporal resolution abilities and the ability to perceive speech in the presence of noise were better in musicians than in non-musicians. The results of temporal modulation transfer function and gap detection threshold values showed that the temporal resolution abilities becomes better as the years of musical experience of the musicians increased. But the results of the speech perception in noise were not statistically significant when the musicians were compared across their experience, though the scores were better in experienced musicians when compared to the musicians with less experience.

Rajalakshmi and Bharath (2012) studied the temporal resolution abilities of children born in musical background and children born in non-musical background. Also, they studied speech perception abilities in the presence of background noise for the same group. A total of 40 subjects participated in the study. An informal questionnaire was administered to all participants, in order to get the information regarding their training in musical field and also familial musical background. The participants were classified into 4 subgroups based on their family background and musical training. Each group consisted of 10 subjects. Temporal resolution abilities were found out using Temporal Modulation Transfer Function (TMTF) and Gap Detection Threshold (GDT) test. Quick Speech Perception in Noise – Kannada (QuickSPIN) was administered to check speech perception in the presence of noise.

The results from the study showed that the temporal resolution abilities and the ability to perceive speech in the presence of noise were better in children with musical training than compared to children without musical training. Musical training as a factor has contributed to better performance whether or otherwise of the family background. Secondly, the family background with musical training has shaped the auditory processing skills to the finest level. In the context of no family background, musical training has yielded good performance. Finally, in the absence of family background and musical training the auditory processing skills have not shown significant differences.

Saha and Rajalakshmi (2013) studied temporal resolution abilities and speech perception in noise in normal hearing mridangam players and non musicians. Temporal modulation transfer function was done in six different modulation frequencies (4 Hz, 8 Hz, 16 Hz, 32 Hz, 64 Hz, & 128 Hz). Three Interval Alternate Forced Choice Method (3IAFC) method was used to obtain the gap detection threshold. Quick Speech-in-Noise – Kannada (Avinash, Methi & Kumar, 2009) was done using 60 sentences based on the subjects rating of predictability. An eight talker speech babble noise was used to generate sentences with different SNRs. In each list first sentence was at +20 dB SNR and SNR was reduced in 5 dB steps for the subsequent sentences. The results indicated that the temporal resolution abilities and the ability to perceive speech in presence of noise were better in musicians than compared to non musicians. But the results of the temporal resolution abilities and the abilities to perceive speech in presence of background noise were statistically not significant when the musicians were compared across their experience.

Zuk et al. (2013) compared musicians and non-musicians on discrimination thresholds of three synthetic speech syllable continua that varied in their spectral and temporal discrimination demands, specifically voice onset time (VOT) and amplitude envelope cues in the temporal domain. Musicians demonstrated superior discrimination for syllables that required resolution of temporal cues.

Frequency Discrimination

Studies investigating frequency discrimination abilities are based on the rationale that musicians are required to detect minute changes for correct tuning of musical instruments and for detecting mistuned melodies. Results of such studies revealed that when no training was provided, thresholds of the musicians were on average two to six times smaller to nonmusicians. After training this difference was reduced to four times smaller (Kishan-Rabin, Amir, Vexler & Zaltz, 2001; Michely, Delhommeau, Perrot & Oxenham, 2006; Spiegel & Watson, 1984).

Categorical Perception

Categorical perception means that a change in some variable along a continuum is perceived, not as gradual but as instances of discrete categories. Categorical perception (CP) is revealed when an observer's ability to make perceptual discriminations between things is better when those things belong to different categories rather than the same category, controlling for the physical difference between the things. CP in speech is the ability to discriminate speech sounds and define those to different phonemic categories (Liberman, Harris, Hoffman & Griffith, 1957). In this process the categories possessed by an observer influence its perception.

CP is an important phenomenon in cognitive science because it represents an essential adaptation of perception to support categorizations that an organism needs to make. It has been argued that the speech processor consists of specialized linguistic feature detectors that are "tuned" to the phonemic distinctions of language (Eimas & Corbit, 1973) and as a result, acoustic variations in the auditory stimulus irrelevant to meaning are filtered out. This gives rise to the phenomenon of categorical perception, a process whereby continuous acoustic variation is transformed into a discrete set of auditory events, the phonemes of a language system. It has been suggested that categorical perception is unique to speech (Liberman, 1970; Liberman, Cooper, Shankweiler & Studdert- Kennedy, 1967; Studdert-Kennedy, Liberman, Harris & Cooper, 1970).

The phenomenon of CP in cognitive science involves the interplay between humans' higher-level conceptual systems and their lower-level perceptual systems. Speech perception requires the effortless mapping from smooth, seemingly continuous changes in the sound features into discrete perceptual units, a conversation exemplified in the phenomenon of categorical perception. CP is theoretical because it offers a potential account for how the apparently symbolic activity of high level cognition can be grounded in perception and action (Harnard, 1990). CP is said to be present when people more reliably distinguish physically different stimuli when the stimuli come from different categories than when they come from the same category (Harnard, 1987).

Categorical perception is an unusual result to obtain in psychophysical experiments with non-speech sounds (Liberman, 1970). /b/ and /p/ are related along the continuum of voice onset time, as are the other voiced and voiceless stops, /g/-/k/ and /d/-/t/ (Lisker & Abramson, 1964). When listeners are presented with synthetic speech sounds varying in voice onset time, they are able to identify the stimuli as examples of the appropriate phoneme class, and there is good agreement among speakers of a language about the transition from a voiced to a voiceless stop. Labeling distributions for speech sounds, particularly in the stop consonants, are regular and symmetrical, with little overlap between adjacent categories, and with steep category boundaries.

A classic demonstration of CP for a certain type of speech stimuli is in terms of its voice onset time (VOT). The VOT of the voiced speech sounds was varied in such a way that it created a continuum ranging from voiced sound at one end of the continuum to voiceless sound at the other end. Voice onset time (VOT) – the time interval between the release from stop closure and the onset of laryngeal pulsing in the following vowel-has been shown to be a sufficient dimension for distinguishing between voiced and voiceless stop consonants in many different languages (Lisker & Abramson, 1964; 1965; 1967). Early studies employing synthetic speech sounds indicated that the voiced- voiceless distinction is perceived in a categorical mode, i.e., the listeners respond to the speech sound as if they set perceptual boundaries around the range of sounds that will be identified as a single phonetic segment.

Speech, it is argued, is handled by a specialized neural processing mechanism localized in the left cerebral hemisphere (Darwin, 1969; Kimura, 1961; Shankweiler & Studdert-Kennedy, 1967), while music is thought to be primarily a right-hemisphere function (e.g., Kimura, 1964; Milner, 1962; Shankweiler, 1966). Music education and training engages all human senses and involves all cognitive processes (sensory, perceptual & cognitive learning, memory, emotion, etc.), but it also requires motor activation (utilized while playing an instrument) and appropriate articulation (utilized while singing or playing). Hence, speech and music which involves the function of both the hemispheres and also in categorical perception, but some functional differences have been observed by Ohnishi and co-workers (2001), who found that there is "a distinct cerebral activity pattern in the auditory association areas and prefrontal cortex of trained musicians".

Musicians to form an ideal subject pool in which one can investigate possible cerebral adaptations to the unique requirements of skilled performance as well as cerebral correlates of unique musical abilities such as absolute pitch and others and hence, the concept of categorical perception. There are several reasons for this. First, the commencement of musical training usually occurs when the brain and its components may still be able to adapt. Second, musicians undergo long-term motor training and continued practice of complicated bimanual motor activity. Third, imaging studies has shown that motor learning and the acquisition of skills can lead to changes in the representation of motor maps and possibly also to micro-structural changes.

Music training hones the perception of minute acoustic differences that distinguish sounds; this training may generalize to speech processing given that adult musicians have enhanced neural differentiation of similar speech syllables compared with non-musicians. Identification ability improves drastically, as in one study we found that highly trained musician could label intervals accurately, reliably, and independently of the stimulus' context (Siegel & Siegel, 1977). Absolute, context-free identification is more usually associated with speech, and is thought to be one of the phenomena that make speech special. Moreover, musicians report hearing intervals as sounding qualitatively different from one another. A major third has a completely different sound from a minor third to a musician. It does not sound 'larger', but rather is described as having a different character-much like the difference between two consonant phonemes. Finally, there is evidence that musicians may exhibit left- hemisphere processing of tonal stimuli (Bever & Chiarello, 1974), just as they do for speech. Non-musicians, on the other hand, seem to show right- hemisphere dominance for music (Kimura, 1964).

ELECTROPHYSIOLOGICAL STUDIES AS EVIDENCE FOR ENHANCEMENT IN AUDITORY PERCEPTION

Auditory neurophysiological measures in musicians both at brainstem and cortical levels have been shown better in musicians.

Parbery-Clark, Skoe and Kraus (2009) compared sub-cortical neurophysiological responses to speech in quiet and in noise in a group of highly trained musicians and non-musician controls. Speech evoked auditory brainstem responses for speech syllable /da/ indicated that musicians exhibited more robust responses in background noise than control group. Musicians also had earlier response onset timing and better phase locking to the temporal waveform and stimulus harmonics than non-musicians. They also found that earlier response onset timing and more robust brainstem responses to speech in background noise were both correlated to better speech in noise perception as measured through HINT. They concluded that musical experience resulted in more robust sub-cortical representation of speech in the presence of background noise, which may contribute to musician's behavioral advantage for speech in noise perception.

Musicians also exhibited more faithful encoding the steady state portion of a stimulus in the presence of background noise. By calculating the degree of similarity between stimulus waveform and the sub-cortical representation of the speech sound, it was found that musicians had higher stimulus-to-response correlations in noise than non-musicians. A greater stimulus to response correlation is indicative of more precise neural transcription of stimulus features.

One possible explanation for this musician enhancement in noise may be based on Hebbian principle, which posits that the associations between neurons that are simultaneously active are strengthened and those that are not are subsequently weakened (Hebb, 1949). It is speculated that extensive musical training may lead to greater neural coherence. This strengthening of the underlying neural circuitry would lead to a better bottom-up; feed forward representation of the signal.

Deepika and Rajalakshmi (2012) compared behavioral scores of speech perception in noise and sub cortical neurophysiological responses to speech in a group of highly trained musicians and non-musician controls. A total of 50 subjects comprised musicians and non musicians were considered, aged between 7-18 years. Speech evoked ABR and LLR were carried out along with SPIN test. Musicians were found to have a more robust sub cortical representation of the speech stimulus. Musicians had earlier response onset timing, than non-musicians. Fo encoding is better in musicians than in non musicians. Speech perception ability in musicians becomes better with increased years of musical exposure and experience. The increase in Fo amplitude is positively correlated with the speech perception in noise. Musicians had better behavioral performance on the Speech in Noise Test (SPIN) outperforming the non-musician controls.

Zubin and Rajalakshmi (2012), conducted a study to investigate the effect of musical training on the encoding of speech stimuli at the level of the brainstem by examining and comparing speech evoked ABR responses of both Carnatic musicians and non-musicians in quiet and in the presence of noise, and by relating these responses to the performance of the subjects in the task of speech perception in noise. 15 musicians and non-musicians between the age ranges of 20 to 50 years participated in that study. They found that, in quiet, the onset peaks V and the transition peaks D, E, and F were clearly visible in the speech evoked ABR of the non-musicians. The morphology of the waves was noticeably poorer in noise, with peaks having reduced amplitude and delayed latencies.

As in Figure 5.1, the V-A complex is almost eliminated in noise, though the transition waves are less affected. Similar findings were reported in studies by Russo, Nicol, Zecker, Hayes and Kraus (2004) & Russo, Nicol, Musacchia and Kraus (2004). In quiet, the morphology of the Speech-Evoked ABR did not vary much from that seen in non-musicians. Though the waveform morphology was poorer in noise than in quiet, the waves were by and large better defined in musicians than in the corresponding waveforms of non-musicians. The V-A complex in particular is more clearly seen (Fig 5.2). This is in line with the findings of Parbery-Clark, Skoe and Kraus (2009).

These are evidences for musical expertise contributing to an enhanced subcortical representation of speech sounds in noise. Musicians had more robust temporal and spectral encoding of the eliciting speech stimulus, thus offsetting the deleterious effects of background noise.

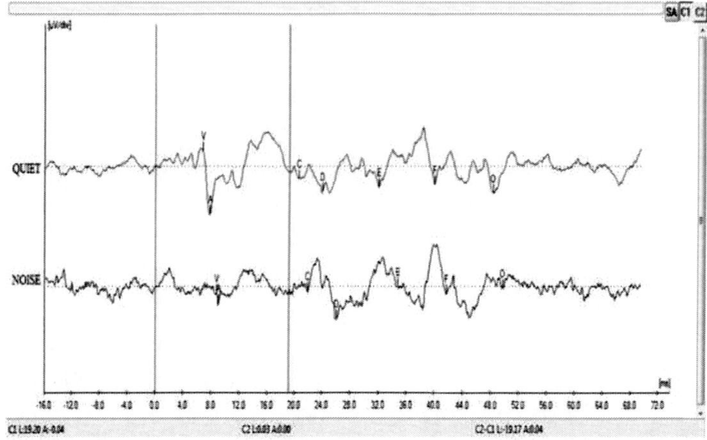

Figure 5.1. Speech-Evoked ABR in response to 40 msec /da/ acquired in a Non-Musician in quiet and in noise (0dB SNR) studied by Zubin and Rajalakshmi (2012).

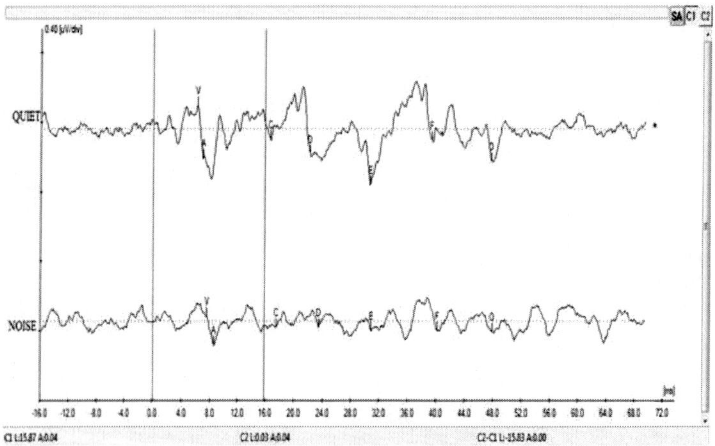

Figure 5.2. Speech-Evoked ABR in response to 40 msec /da/ acquired in a Musician in quiet and in noise (0dB SNR), studied by Zubin and Rajalakshmi (2012).

Faster neural timing and enhanced harmonic encoding in musicians suggests that musical experience confers an advantage resulting in more precise neural synchrony in the auditory system. These findings provide a biological explanation for musicians' perceptual enhancement for speech-in-noise (Anderson & Kraus 2010).

Musacchia, Strait and Kraus (2008) studied the relationship between evoked potentials and musical experience. They recorded simultaneous brainstem and cortical evoked potentials in musicians and non-musician controls. Because previous research showed that musician related effects extend to speech and multi-sensory stimuli, the speech syllable /da/ was presented in three conditions: when subjects listened to auditory sound alone, when the subjects simultaneously watched a video of a male speaker saying /da/ and when they viewed the video alone. The analysis focused on comparing measures of the speech evoked brainstem response that have been previously reported as enhanced in musicians with well established measurements of cortical activity (e.g., P1-N1-P2 complex).

The first picture that emerged from the data was that recent musical training improves one's auditory memory and shapes composite (P1-N1) and pitch encoding (F0) in a co-coordinated manner. The evoked potentials and behavior correlations suggest that complex auditory task performance is related to the strength of the P1-N1 response. The instrumental musicians performed better in the behavioral tests and had steeper P1-N1 slopes than non-musicians. However, it was not only the individual tests and measures that were music related. Musicians had a statistically stronger correlation between this set of brain and behavior measures than non-musicians. While it is well known that trained musicians outperform untrained controls and have more robust evoked-potentials than non-musicians, the present data showed that the accord, or relationship, between brain and behavior is also improved in musicians.

Shahin, Roberts and Trainor (2003) compared AEPs evoked by pure, violin and piano tones in young 4- to 5- year old children with age matched non-musician children. The aim of the study was to assess whether AEP components are sensitive to musical experience at this age and, if so, which components are affected. Larger amplitude P1, N1, and P2 responses were found in 4-to 5-year-old musically experienced children compared with musically less experienced children. Furthermore, the P2 enhancement was specific to the instrument of practice. Thus, they concluded that, AEPs differ between musical and control children as young as 4 years of age, and the differences reflect specific musical experience. Comparison of piano-evoked N1 and P2 responses in those 4- to 5-year-old musicians (most of whom were pianists) to cross sectional findings suggest that musical experience may have advanced the developmental trajectory for sounds of the instrument of training.

Koelsch et al. (1999) and Russeler et al. (2001) compared musicians and non-musicians by using mismatch negativity (MMN), a component of the evoked potentials in auditory cortex, obtained by submitting subjects to hundreds of identical stimuli that are randomly replaced by distinct ones, in the absence of attention to stimulus. Koelsch et al. observed MMN for slightly impure chords presented among perfect major chords, only in musicians, suggesting that they are superior in pre-attentively extracting more information out of musically relevant stimuli. Russeler et al. demonstrated that while musicians showed MMN for regularly spaced tones mistimed by 20 milliseconds, presence of MMN in non-musicians needed mistiming of longer than 50 milliseconds, indicating that temporal integration may be more precise in musicians.

Lappe et al. (2008) showed that even short-term musical training in adults can induce cortical plasticity. In their study, a group of adults learned to play a piano sequence, while the control group just listened to the music and judged it. Their results indicated that the group that actively played piano had an enlargement of Mismatch Negativity Potentials after training, while passive listeners did not show such a pattern.

In sequential stimuli, a wrong note occurring occasionally in a short melody that is repeated in different keys (i.e., starting on different notes) from trial to trial, elicit frontally negative event-related potential i.e. MMN. While MMN to such melodic changes is present in both musicians and non-musicians, it is much larger in musicians (Fujioka, Trainor Ross, & Kakigi, 2004). In terms of polyphonic music, altered notes in either of simultaneous melodies elicit MMN responses that were found to be larger in musicians than non-musicians (Fujioka, Trainor, Ross, Kakigi & Pantev, 2005).

In vocal and instrumental musicians pre-attentive auditory discrimination has been assessed by auditory evoked potentials, considering mismatch negativity (Nikjeh & Adams, 2006), they reported superior mismatch negativity responses in both vocal and instrumental musicians compared to non musicians.

It is very well established that musicians spend years fine-tuning their skills on music which is a complex auditory task. Research has accumulated lot of evidence supporting how their fine tuning skills lead to superior performance in auditory processing when compared with non musicians. With the above examples it is clear that, behavioural have offered evidence for enhancement in auditory perception measured through temporal processing abilities, frequency discrimination, categorical perception and speech perception in noise. Similarly electrophysiological evidences supporting the superior performance in musicians range from sub-cortical neurophysiological responses to speech in quiet and in noise and more faithful encoding of the cortical potentials like LLR and MMN. These scientific proof thus evidenced makes one to believe that the there are benefits of musical training on perceptual domain.

REFERENCES

Alain, C., & Brenstein, L. J. (2008). From sound to meaning: The role of attention during auditory scene analysis. *Current Opinions of Otolaryngologica: Head and Neck Surgery, 16,* 485-489.

Anderson, S., Kraus., N., (2010). Sensory-cognitive interaction in the neural encoding of speech in noise: A review (2010). Journal of American Academy of Audiology 21, 575-585.

Anderson, S., Parbery-Clark, A., Yi, H.,&Kraus., N., (2011). A neural Basis of speech-in-noise perception in older adults. *Ear & Hearing, 32,* 1-8.

Anderson, S., Skoe, E., Chandrasekaran, B,. Zecker, S., & Kraus, N,.(2011) Brainstem Correlates of Speech-in-Noise Perception in Children. *Hearing Research ,270,*151-157.

Assmann, P. F., & Summerfield, Q. (1990).Modeling the perception of concurrent vowels: Vowels with different fundamental frequencies. *Journal of Acoustical Society of America, 88,* 680-697.

Beavois, M. W., & Meddis, R. (1997).Time decay of auditory stream biasing. *Perception and Psychophysics, 59,* 81-86.

Besson, M., Schon, D., Moreno, S., Santos, A., & Magne, C. (2007). Influence of musical expertise and musical training on pitch processing in music and language. *Restorative Neurology and Neuroscience, 25,* 399-410.

Bialystok, M. W., & DePape, A. M. (2009). Musical expertise, bilingualism, and executive functioning. Journal of Experimental Psychology: *Human Perception and Performance, 35,* 565-574.

Boersma, P. & Weenick, D. (2005).*Praat doing phonetics by computer* (version 4.6.09) from www.praat.org

Bradlow, A. R., Kraus, N., Hayes, E. (2003). Speaking clearly for children with learning disabilities: sentence perception in noise. *Journal of Speech Language and Hearing Research, 46,* 80-97.

Brandler, S., & Rammsayer, T. H. (2003).Differences in mental abilities between musicians and non-musicians. *Psychology of Music, 31 (2)*, 123-138.

Brandt, J., Rosen, J.J. (1980). Auditory phonemic perception in dyslexia: categorical identification and discrimination of stop consonants. *Brain and language, 9*, 324-337.

Brattico, E., Naatanen, R., & Tervaniemi, M. (2001). Context effects on pitch perception in musicians and nonmusicians: Evidence for event related potentials. *Music Perception, 19*, 199-222.

Bregman, A. S. (1990).*Auditory Scene Analysis: The Perceptual Organization of Sound*. Cambridge, MA: MIT Press.

Brokx, J. P. L., & Nooteboom, S. G. (1982).Intonation and the perceptual separation of simultaneous voice. *Journal of Phonetics, 10*, 23-36.

Chan, A. S., Ho, Y. C., & Cheung, M. C. (1998). Music training improves verbal memory. *Nature, 396*, 128.

Chandrasekaran, B., Krishnan, A., & Gandour, J. T. (2009). Relative influence of musical and linguistic experience on early cortical processing of pitch contours. *Brain and Language, 108*, 1-9.

Chandrasekaran., B., Kraus., N., (2010). The scalp-recorded brainstem response to speech: Neural origins and plasticity. *Psychophysiology*.47, 236–246.

Deepika, V., & Rajalakshmi, K. (2012).*Auditory plasticity in musicians: A comparative study*. Published master's dissertation, submitted to University of Mysore.

Elbert, T., Pantev, C., Wienbruch, C., Rockstroh, B., Taub, E. (1995).Increased cortical representation of the fingers of the left hand in string players. *Science, 270 (5)*, 305-307.

Foregeard, M., Schlaug, G., Norton, A., Rosam, C., Iyengar, U., & Winner, E. (2008).The relation between music and phonological processing in normal-reading children and children with dyslexia. *Music Perception, 25*, 383-390.

Foregeard, M., Winner, E., Norton, A., & Schlaug, G. (2008). Practicing a musical instrument in childhood is associated with enhanced verbal ability and nonverbal reasoning. *PLoS One, 3*, 35-66.

Fujioka, T., Trainor, L., & Ross, B. (2008). Simultaneous pitches are encoded separately in auditory cortex: An MMN study. *Neuroreport, 19*, 361-368.

Fujioka, T., Trainor, L., Ross, B., Kakigi, R., & Pantev, C. (2004). Musical training enhances automatic encoding of melodic contour and interval structure. *Journal of Cognitive neuroscience, 16(6)*, 1010-1021.

Fujioka, T., Trainor, L., Ross, B., Kakigi, R., (2004).Musical Training Enhances Automatic encoding of Melodic Contour and Interval Structure, *Journal of cognitive Neuroscience* 16, 1010-1021.

Gaser, C., Schlaug, G. (2003). Brain structures differ between musicians and non-musicians. *Journal of Neurosciences, 23 (27)*, 9240-9245.

Gordon-Salant, S., Fitzgibbons, P. J. (1995).Recognition of multiply degraded speech by young and elderly listeners. *Journal of Speech Language and Hearing Research, 38*, 1150-1156.

Hebb, D.O.(1949).*The Organization of Behavior*. New York: Wiley & Sons.

Ho, Y.C., Cheung, H. C., & Chan, A. S. (2003). Music training improves verbal but not visual memory: Cross sectional and longitudinal explorations in children. *Neuropsychology, 17*, 439-450.

Hoormann, J., Falkenstein, M., Hohnsbein, J., & Blanke, L. (1992). The human frequency following response (FFR): normal variability and relation to the click-evoked brainstem response. *Hearing Research, 59 (2)*, 179-188.

Ishll, C., Arashiro, A. M., & Pereira, L. D. (2006).Ordering and temporal resolution professional singers and amateurs tuned and un-tuned. *Pro-Fono, 18(3)*, 285-292.

Jackobsen, L. S., Cuddy, L. L., & Kilgour, A. R. (2003).Time tagging: A key to musicians' superior memory. *Music Perception, 20*, 307-313.

Jakobson, L. S., Lewycky, S.T., Kilgour, A. R., & Stoesz, B. M. (2008). Memory for verbal and visual material in highly trained musicians. *Music Perception, 26*, 41-55.

Janet, D., & Yathiraj, A. (2003).Effect of Music Training on Auditory Perceptual Skills. Unpublished Master's Dissertation, University of Mysore.

Jeon, J. Y., & Fricke, F. R. (1997).Duration of perceived and performed sounds. *Psychology of Music, 25*, 70–83.

Koelsch, S., & Mulder, J., (2002).Electric brain responses to inappropriate harmonies during listening to expressive music. *Clinical Neurophysiology, 113*, 862-869.

Koelsch, S., Schroger, E., & Tervaniemi, M. (1999).Superior attentive and pre-attentive auditory processing in musicians. *Neuroreport, 10*, 1309–1313.

Koelsch, S., Schulze, K., & Sammler, D. (2008). Functional architecture of verbal and tonal working memory: An fMRI study. *Human Brain Mapping, 30*, 859-873.

Koffka, K. (1935). Principles of Gestalt Psychology. New York: Harcourt, Brace and World.

Kraus, N., & Banai, K. (2007). Auditory-processing malleability: Focus on language and music. *Current Directory of Psychological Sciences, 16,* 105-110.

Kraus, N., Skoe, E., Parbery-Clark, A., & Ashley, R. (2009). Experience induced malleability in neural encoding of pitch, timbre and timing: implications for language and music. *Annals of New York Academy of Sciences 1169,* 543-557.

Krishnan, A., Xu, Y., Gandour, J., Cariani, P. (2005).Encoding of pitch in the human brainstem sensitive to language experience. *Brain Research Cognitive Brain Research, 25 (1),* 161-168.

Kuriki, S., Kanda, S., & Hirata, Y. (2006).Effects of musical experience on different components of MEG responses elicited by sequential piano-tones and chords. *Journal of Neuroscience, 26,* 4046–4053.

Lee, K. M., Skoe, E., Kraus, N., & Ashley, R. (2009).Selective subcortical enhancement of musical intervals in musicians. *Journal of Neurosciences, 29,* 5832-5840.

Levit, H. (1971). Transformed Up-Down methods in psychoacoustics, *Journal of Acoustical Society of America, 49,* 467-477.

Magne, C., Schon, D., & Besson, M. (2006). Musician children detect pitch violations in music and language better than non-musician children: Behavioural and electro-physiological approaches. *Journal of Cognitive Neurosciences, 18,* 199-211.

Marques, C., Moreno, S., & Casstro, S. L. (2007). Musicians detect pitch violation in a foreign language better than nonmusicians: Behavioural and electrophysiological evidence. *Journal of Cognition and Neuroscience, 19,* 1453–1463.

Menning, H., Roberts, L., Pantev, C., (2000). Plastic changes in the auditory cortex induced by intensive frequency discrimination training. *Auditory and Vestibular Systems* 11, 817- 822.

Micheyl, C., Delhommeau, K., Perrot, X., & Oxenham, A. J. (2006). Influence of musical and psycho acoustical training on pitch discrimination. *Hearing Research, 219,* 36–47.

Monteiro, R. A., Nascimento, F. M., Soares, C. D., &Ferreira, M. D. (2010).Temporal resolution abilities in Musicians and no musicians. *International Archives of Otolaryngology, 14(3),* 302-308.

Moreno, S., Marques, C., Santos, A., Santos, M., Castro, S.L., & Besson, M. (2009). Musical training influences linguistic abilities in 8-year old children: more evidence for brain plasticity. *Cerebral Cortex 19*, 712-723.

Musacchia, G., Sams, M., Skoe, E., & Kraus, N. (2007). Musicians have enhanced subcortical auditory and audiovisual processing of speech and music. *Proceedings of the National Academy of Sciences of United States of America, 104*, 15894-15898.

Nair, D., Large, W. E., Steinberg, F., & Kelso, J. A. S. (2002). Expressive timing and perception of emotion in music: an fMRI study. *Proceeding of the 7th International Conference on Music Perception and Cognition, 627*, 627-630.

Ohinshi, T., Matsuda, H., Asada, T., Aruga, M., Hirakata, M., Nishikawa, M. (2001).*Cerebral Cortex, 11(8)*, 75-760.

Oxenham, A. J., Fligor, B. J., Mason, C. R., & Kidd, G. (2003).Informational masking and musical training. *Journal of Acoustical Society of America, 114*, 1543–1549.

Pantev, C., Oostenveld, R., Engelien, A., Ross, B., Roberts, L.E., Hoke, M. (1998). Increased auditory cortical representation in musicians. *Nature, 392*, 811–814.

Pantev, C., Roberts, L. E., Schulz, M., Engelien, A., Almut., Ross.,& Bernhard. (2001). Timbre-specific enhancement of auditory cortical representations in musicians. *Neuroreport, 12*, 169–174.

Parbery-Clark, A., Skoe, E., & Kraus, N. (2009). Musical experience limits the degradative effects of background noise on the neural processing of sound. *Journal of Neuroscience, 29*, 14100-14107.

Parbery-Clark, A., Skoe, E., Lam, C., & Kraus, N. (2009). Musical experience limits the degradative effects of background noise on the neural processing of sound. *Journal of Neuroscience, 29,* 14100-14107.

Parbery-Clark, A., Skoe, E., Lam, C., & Kraus, N. (2009).Musician enhancement for speech in noise. *Ear and Hearing. 30*, 653-661.

Patel, A. D. (2003). Language, music, syntax and the brain. *Natural Neuroscience, 6*, 674-681.

Patel, A. D. (2007). *Music, Language and the Brain*. Oxford: Oxford University Press.

Rajalakshmi, K., & Bhushan, B. (2012).Comparison of temporal resolution abilities and speech perception in noise in children born in families with musical background and children born in families without musical background. Unpublished department project in department of Audiology, AIISH, Mysore.

Rammsayer, T., & Altenmuller, E. (2006).Temporal information processing in musicians and non-musicians. *Music Perception, 24,* 37-48.

Russo N, Nicol T, Zecker S, Hayes E, Kraus N. (2005)Auditory training improves neural timingin in the human brainstem. *Behavioural Brain Research*156: 95-103.

Saha, A., & Rajalakshmi. K. (2013).*Temporal resolution and speech perception abilities in noise in mridangam players.* Unpublished masters dissertation, submitted to University of Mysore, Mysore.

Samelli, A. J., & Schochat, E. (2008). The gaps-in-noise test: Gap detection thresholds in normal-hearing young adults. *International Journal of Audiology, 47,* 238-245.

Schalaug, G. (2001). The brain of musicians: a model for functional and structural adaptation. *Annals of New York Academy of Sciences, 930,* 281-299.

Scheffers, M., (1983).*Sifting Vowels: Auditory Pitch Analysis and Sound Segregation.* Unpublished Doctoral dissertation. Groningen. The Netherlands: Groningen University.

Schlaug, G. (2001). The brain of musicians: a model for functional and structural adaptation. *Annals of New York Academy of Sciences, 930,* 281-299.

Schlaug, G., Norton, A., Overy, K., & Winner, E. (2005).Effects of music training on the child's brain and cognitive development. *Annals of New York Academy of Sciences, 930,* 281-299.

Schneider, P., Scherg, M., Dosch, H. G., et al. (2002). Morphology of Heschl's gyrus reflects enhanced activation in the auditory cortex of musicians. *Natural Neuroscience, 5,* 688-694.

Schon, D., Magne, C., & Besson, M. (2004). The music of speech: Music training facilitates pitch processing in both music and language. *Psychophysiology, 41,* 341–349.

Shahin, A., Bosnyak, D., Trainor, L., Roberts, Larrey, R., (2003).Enhancement of neurolplastic P2 and N1c auditory evoked potentials in musicians. The Journal of Neuroscience, 12, 5545–5552 .

Shahin, A. J., Roberts, L. E., & Pantev, C. (2007).Enhanced anterior-temporal processing for complex tones in musicians. *Clinical Neurophysiology, 118,* 209–220.

Strait, D., Kraus, N., Parbery-Clark, A., & Ashley, R. (2010). Musical experience shapes top-down auditory mechanisms: Evidence from masking and auditory attention performance. *Hearing Research, 261,* 22-29.

Tervaniemi, M., Just, V., Koelsch, S., Widmann, A., Schorger, E. (2005). Pitch discrimination accuracy in musicians vs. non-musicians: an event-related potential and behavioural study. *Experimental Brain Research, 161*, 1- 10.

Thomas, A. O., & Rajalakshmi, K. (2011).*Effect of musical training on temporal resolution abilities and speech perception in noise.* Published Master's Dissertation, submitted to University of Mysore, Mysore.

Thompson, W. F., Schellenberg, E. G., & Husain, G. (2004). Decoding speech prosody: Do music lessons help? *Emotion, 4,* 46-64.

Treisman, A. M. (1964). Verbal cues, language and meaning in selective attention. *American Journal of Psychology, 77,* 206-219.

Tremblay, K., Kraus, N., McGee, T., Ponton, C., & Otis B. (2001). Central auditory plasticity: changes in the N1-P2 complex afte speech-sound training. *Ear & Hearing,* 22, 79–90.

Tzounopoulus, T., Kraus, N., Learning to encode timing : Mechanisms of plasticity in auditory brainstem. *Neuron* 62, 463-469.

Wong, P. C., & Perrachione, T. K. (2007). Learning pitch patterns in lexical identification by native English speaking adults. *Applied Psycholinguistics, 28,* 565-585.

Wong, P. C., Skoe, E., Russo, N. M., Dees, T., & Kraus, N. (2007). Musical experience shapes human brainstem encoding of linguistic pitch patterns. *Nature Neuroscience, 10,* 420-422.

Yee, W., Holleran, S., & Jones, M. R. (1994). Sensitivity to event timing in regular and irregular sequences: Influences of musical skill. *Perception & Psychopsysics, 56,* 461-471.

Zendel, B. R., Alain, C. (2009). Concurrent sound segregation is enhanced in musicians. *Journal of Cognitive Neuroscience, 21,* 1488-1498.

Zendel, B. R., Alain, C. (2009). Concurrent sound segregation is enhanced in musicians. *Journal of Cognitive Neuroscience, 21,* 1488-1498.

Zhang, Y., & Suga, N. (1997).Corticofugal amplification of subcortical responses to single tone stimuli in the mustached bat. *Journal of Neurophysiology, 78,* 3489-3492.

Ziegler, J. C., Pech-Georgel, C., George, F., Alario, F. X., Lorenzi, C. (2005). Deficits in speech perception predict language learning impairment. *Proceedings of National Academic Science USA, 102,* 14110-14115.

Zubin, V., & Rajalakshmi, K. (2012). Brainstem correlates of speech perception in noise: Carnatic Musicians versus Non-Musicians, Published Master's Dissertation, submitted to University of Mysore, Mysore.

Chapter 6

Music Induced Hearing Loss

It has been generally accepted that excessive exposure to loud sounds/music causes various hearing loss symptoms (e.g. tinnitus) and consequently leads to a risk of permanent hearing damage. Such potential risk of hearing damage due to loud music exposure has been widely investigated in musicians and people working in music venues. With advancements in sound technology and rapid developments in the music industry, increasing number of people, particularly adolescents and young adults, are exposing themselves to music on a voluntary basis at potentially harmful levels, and over a substantial period of time, which can also cause hearing loss. Rock musicians and classical musicians are at higher risk of music induced hearing loss. The risk of damage to hearing when listening to music depends on factors such as how loud the music is; how close one may be to speakers; how long and how often one is exposed to loud music; headphone use; family history of hearing loss. Jobs or activities that increase the chance of music induced hearing loss are: being a musician, sound crew member, or recording engineer; working at a night club; attending concerts; using portable music devices with headphones.

Negative Influence of Music Exposure/Training on the Auditory System

The importance of hearing acuity for musicians cannot be overstated. Unfortunately, the levels of exposure during practice, rehearsal and performance are capable of damaging the hearing mechanism. Musicians are especially vulnerable to the effects of high noise levels due to their continual exposure in practice and performance (Early & Horstman, 1996). Though their exposure is intermittent, musicians still are at risk for suffering from the devastating effects of Music Induced Hearing Loss (MIHL) because of the sound levels produced by instruments.

Evidence has demonstrated that exposure to non-amplified and low intensity sound over a period of time has a positive impact on musicians, known as the 'protective mechanism of sound conditioning' against destructive effects of noise trauma (Niu & Canlon, 2002; Miyakita, Hellström, Frimansson & Axelsson, 1992). Such a protective mechanism of the auditory system may be limited, depending on the frequency range exposed, intensity and duration. In the study by Kazkayasi et al. (2006), although their results indicated musical training had positive effects in terms of hearing acuity and musical perception using conventional audiometric measurements, they also found hearing reduction in extended high frequencies of 12, 14, and 16 kHz after two years of musical training and practice. The decrease in average hearing acuity at these frequencies might be attributed to continuous music exposure.

There is ongoing debate regarding a correlation between high sound levels and hearing disorders. The most commonly assumed causes of hearing disorders that affect musicians are the high sound levels to which they are exposed (Hart, Geltman, Schupbach & Santucci, 1987; Drake-Lee, 1992; Early & Hortsmann, 1996). These hearing disorders can manifest themselves in several ways, and can represent a great burden for those affected. From the previous studies done on classical musicians, five different hearing disorders were assumed to be caused by high sound levels resulting from music. They are: Hearing loss, tinnitus, hyperacusis, distortion, diplacusis (Kahari, Axelsson Hellstrom & Zachau, 2001). However, controversies are reported justifying the fact that music is more intermittent and varying having less detrimental effect on hearing acuity.

Thus, few studies report that music is more pleasant and does not exhibit oto-traumatic effect on musicians. On the other hand there are various studies which report musicians are at risk to develop music induced hearing loss due to continual exposure to loud music.

Studies which Report Unaffected Hearing with High Levels of Musical Exposure

Karlsson et al. (1983) studied 417 symphony orchestra musicians wherein the median values of age groups were compared with normative reference It was found that there was no difference between the measured PTA in these musicians when compared with that of normative reference. Johnson et al. (1986) examined the hearing acuity among orchestral musicians compared to non-musicians. Using conventional audiometry (250-8K Hz) and EHF (9- 20K Hz) obtained thresholds were examined comparing musician versus non-musician, age and gender. It was found that the hearing acuity among the musicians and non-musicians were similar and a consistent decrease in hearing acuity as the frequency increased. The authors concluded that there was no prominent hearing threshold decline due to consistent noise exposure among musicians as well as no apparent gender or inter-aural differences.

Deatherage (2003) studied the extended high frequency hearing of 66 subjects, aged 18-27 years. Subjects participated in the study included 33 musicians and 33 non-musicians. Comparisons between the extended high frequency thresholds (EHF) and conventional pure tone thresholds between musicians and non-musicians were made. The results revealed that there was no statistically significant difference in auditory thresholds between musician and non-musician group. The results however illustrated slightly better thresholds among the musicians than non-musicians in the EHF, though not statistically significant.

Studies Which Report Affected Hearing with High Musical Exposure

Axelsson and Lindgren (1981) studied 139 classical musicians (122 males and 17 females) and found that 80 (58%) musicians were identified as having hearing loss and among them 51 (37%) had hearing loss being partially or wholly due to music exposure. There was one more study in the year 1981 conducted by Westmore and Eversden wherein 34 orchestral musicians were considered in the study. 23 out of 68 ears (34%) showed audiogram pattern consistent with NIHL and 4 out of those 23 ears had a hearing loss more than 20 dB at 4 kHz.

Ostri et al. (1989) studied 96 orchestral musicians (80 males and 16 females; aged between 22 and 64 years). In this study the PTA was compared to normative reference (ISO 7029) and it was found that 58% of musicians had hearing loss of which 50% of males & 13% of females showed typical audiograms of NIHL. Rajalakshmi & Apoorva (2013) have reported that there was significantly poor hearing thresholds on left ear at higher frequencies in violinists. Another study which compared the PTA with normative reference (ISO 7029) showed that the audiograms in 52.5% of ears suggested NIHL (Royster, Royster & Killion 1991). Royster et al. also reported significant poorer thresholds in the left ear when compared to right ear in violinists.

Mc.Bride, Gill, Proops, Harrington, Gardiner and Attwell (1992) examined 89 classical orchestral musicians by comparing the hearing levels between 18 woodwind and brass Musicians (high risk group) with 18 string musicians (low risk group) matched for age and sex. It was found that there was no significant difference in hearing thresholds between high risk group and low risk. The authors also confirmed that noise levels for classical musicians are not only considered uncomfortable but exceed permissible occupational noise level standards. Another study revealed that after exposure to unamplified music 31% of young musicians had a TTS of 15-20 dB. Approximately 50% of the subjects who had been exposed to amplified music had a TTS of 15-20 dB. Usually the TTS was measured at 6K Hz and the sources of the highest noise levels were percussion, brass and loud speakers (Fearn, 1993).

Kahari et al. (2001) compared 140 classical orchestral musicians (98 males and 42 females; aged between 23 and 64 years; mean age = 40 years) with normative reference (ISO 7029). The authors reported that the hearing thresholds for female musicians were significantly better than male musicians in high frequency and the median audiogram for male musicians displayed high frequency notch at 6 kHz. Kahari, Zachau, Eklof, Sandsjo and Moller (2003) assessed hearing and hearing disorders among 139 rock/jazz musicians and found that around 74% of them had some or the other form of hearing disorders. Hearing loss, tinnitus and hyperacusis were most commonly seen hearing disorders. The 3-6 k Hz notch was shown and the woman showed bilateral significantly better hearing thresholds at 3-6 kHz than the men.

In 2006, Schmuziger, Patscheke and Probst studied 42 rock and roll musicians for whom PTA (0.25 to 14 kHz) was administered. Assessments of tinnitus and hypersensitivity to sounds were also carried out. The results of the study revealed significantly elevated hearing thresholds averaged at 3 to 8 kHz in musician group than the control group. 11 of the musicians (26%) were found to be hypersensitive to sound, and 7 (17%) presented with tinnitus.

Maia and Russo (2008) studied 23 rock and roll musicians. PTA, Tympanometry and OAE tests were used in the present study. The findings of the study showed that 100% of the ears presented thresholds within normal limits; however, 41% of the audiograms had notches at frequencies between 4 and 6 kHz. The results also revealed Reduced OAE amplitudes recorded in musicians (61% absent OAEs) even when the hearing loss was not found. The authors also reported that the other hearing symptoms reported by the musicians were intolerance for intense sounds (48%), tinnitus (39%) and auricular fullness (22%).

Classical orchestra musicians (109) were studied by Emmerich et al. (2008). PTA (0.25 to 16 kHz) and OAE tests were carried out in the study. The PTA criteria considered was the occurrence of permanent threshold shift (PTS) larger than 15 dB Noise levels which is generally reported to be higher than the regulated standards. The results of the study found that more than 50% of the musicians were found to have permanent hearing shift of 15 dB or more and there was a significant decline in OAE amplitudes which were correlated with the length of time of being professional musicians.

From the above findings it can be noted that music training/ exposure has both positive and negative influence on the auditory system. The above studies were related to audiological findings in musicians to investigate the effect of music training/exposure on the auditory system. There are also various studies related to sound level measurements of different musical instruments and vocal input which in turn reflect on probability of hearing damage risk in the musicians.

Studies on Sound Level Measurements of Vocal Output and Output of Various Musical Instruments

Before actually knowing about the sound output of different instruments and whether they are harmful or not, let us get aware about what is the maximum exposure limits for sound. Table 6.1 depicts the exposure limits as per National institute for Occupational Safety and Health (NIOSH) and Occupational Health and Safety administration.

Table 6.1. Duration in hours of allowable exposure
as per NIOSH and OSHA

Sound level in dB A	Duration of allowable exposure (Hours)	
	NIOSH	OSHA
85	16	8
88		4
90	8	
92		
94		1
95	4	
100	2	0.25
105	1	
110	0.5	
115	0.25	

According to this Criteria getting exposed to sounds greater than 90 dB for more than 8 hours is going to be harmful for our ears (NIOSH). Also, when sound levels are as high as 105 dB A, listening to them even for more than an hour can cause damage. Folprechtov and Mikxovsk (1976) carried out sound level measurements of different musical instruments and measured sound levels of 92 dB (A) with variations of 87-98 dB (A) in a symphony orchestra. Usually the musicians performed 4-8 hours a day. The sound levels of the different instruments were as follows:

Table 6.2. Sound levels outputs of different musical instruments studied
by Folprechtov and Mikxovsk (1976)

Instrument	Sound output
Violin	84 - 103 dB(A)
Cello	84 - 92 dB (A)
Base	75 - 83 dB (A)
Piccolo	95 - 112 dB (A)
Flute	111 dB (A)
French horn	90 - 106 dB (A)
Oboe	80 - 94 dB (A)
Trombone	85 - 114 dB (A)
Xylophone	90 - 92 dB (A)
Clarinet	92 -103 dB (A)

Royster et al. (1991) assessed the risk of NIHL among symphony orchestra musicians using sound level measurements. Personal dosimeter set to the 3 dB exchange rate was used to measure sound levels during rehearsals and concerts. The Leq values ranged from 79 to 99 dB (A), with a mean of 89.9 dB (A).

An assessment of noise exposure and hearing thresholds was carried out by Backus, Clark and Williamon (2007) in orchestral musicians during rehearsals and performance. The authors measured maximum personal daily noise exposure levels and it was found that the musicians were exposed to sound levels above 85 dB (A). The exposure levels depended strongly on the instrument being played and where the musicians were seated in the orchestra.

In a study by Philips and Mace in 2008, which was done with the primary objective of determining sound levels in student practice rooms, average sound levels and percentage of daily dose of noise exposure were measured. Also the authors determined whether any instrument group was at higher risk for music-induced hearing loss due to exposure levels. Brass, wind, string and voice instruments were measured in the study. Measurements were taken using a dosimeter or Dose Badge clipped to the shoulder during 40 students' individual practice sessions. Mean sound levels measured averaged 87-95 dB (A). Mean average levels for the brass players were significantly higher than other instrument groups.

Exposure to sound and the risk of noise-induced hearing loss in orchestral musicians was assessed by Pawlaczyk-Łuszczyńska, Dudarewicz, Zamojska, & Śliwińska-Kowalska in 2010. Sound level measurements were carried out within one opera and two symphony orchestras along with questionnaire inquiries in their employees. Based on this data, the risk of developing noise induced hearing loss was assessed according to ISO 1999:1990. The authors reported that the classical orchestral musicians are usually exposed to sound at equivalent continuous A-weighted sound pressure levels of 79−90 dB, for 20−45 hours per week. They also reported that exposures to such high sound levels over 40 years of employment might cause the risk of hearing impairment in the range of 4−30% and 16−43% in case of females and males, respectively. The authors measured the sound levels of different musical instruments and it was found that the highest risk is related to playing a clarinet (up to 35%), tube (up to 35%), trombone (up to 35%), trumpet (up to 40%), percussion section (up to 41%) and horn (up to 43%).

Rajalakshmi and Apoorva (2012) in their study with a total of 40 participants in the ages of 30-55 years were classified into 2 groups. *Experimental Group:* Professional musicians who were involved in teaching music served as subjects for the study. Both vocal and instrumental musicians were considered. 5 vocal musicians and among instrumental music, 5 violinists (String), 5 Mridangam players (percussion) and 5 flutists (wind) participated. A total of 20 musicians participated in the study. *Control Group:* 20 age matched individuals with non-musical background served as controls.

The testing was carried out in 2 phases. Phase I: Audiological profiling of musicians and non-musicians. Phase II: Sound level measurements of professional musicians during teaching. During Phase I, all participants underwent a questionnaire interview. Questionnaire used included 5 domains - Basic information, Musical history, Medical history, Life-style and Self-assessment of hearing status. The questionnaire was administered on both the groups (musicians and non-musicians). Information sought from musical history included queries on musical training and musical proficiency, regarding their musical performances/Concerts and on music teaching sessions and exposure to music during teaching hours. Interview was followed by Otoscopy, pure-tone audiometry, extended high frequency audiometry, speech audiometry, immittance audiometry and OAE measurements. To avoid effects of TTS all the evaluations were done after providing a hearing rest (> 8 hours without music exposure).

Phase II involved measuring sound levels during professional musicians involved in teaching: The position of the microphone was placed at the ear level of the musician at 1 foot distance and at 45 degree azimuth. 60 – 100 SLM readings were taken over a period of 15-20 min (4 readings in a minute) and the Leq and Lmax were calculated. Results showed no significant difference in auditory thresholds between each of the musician group and non-musician group. It was found that there was no difference in SRT and SIS between the groups. However LDLs were found to be higher in musicians when compared to non-musicians and statistically significant difference was observed for flutist and mridangists. A similar finding was found in a study done by Axelsson and Lindgren, 1981. They reported that musicians may be accustomed to wide dynamic ranges with daily exposures to sounds of varied intensities. Thus, higher LDLs observed for musician group can be attributed to the higher tolerance levels developed by them due to frequent exposure to musical sounds. Further, the two groups did not show any significant difference for Acoustic Reflex thresholds and OAE amplitudes.

The Leq and Lmax mean and standard deviation values are shown in the table 6.3. It can be noted that the flutists had the maximum sound exposure while teaching. Leq and Lmax values from highest to lowest is as follows - Flutists - Mridangists - Vocalists - Violinists. Thus, flutists are at the highest risk to develop hearing damage because of the loud music exposure. A similar finding was reported in a previous study by Folprechtov and Mikxovsk (1976) wherein the highest sound output was reported for flute - 111 dB (A).

Table 6.3. Mean Leq and Lmax values for different musician sub- groups studied by Rajalakshmi and Apoorva (2012)

Musician groups	Leq	Lmax
	Mean (dB)	Mean (dB)
Vocalists	91.83	97.45
Mridangists	92.21	101.54
Flutists	101.54	107.3
Violinists	85.94	95.16

Factors That Increase the Risk Hearing Problems in Musicians

There are several risk factors for noise-induced hearing loss (NIHL) in musicians of all types. Some are quite obvious such as playing music that are consistently loud (Chesky and Henoch 2000; Kahari, Zachau et al. 2004), or playing music with high intensity at high frequencies (as in some symphonies). Perhaps equally obvious is the observation that being exposed for longer periods of time, whether over a period of one day or over a lifetime, poses a higher risk for developing hearing loss (Kahari, Axelsson et al. 2001). Playing some of the "louder" instruments (such as bassoon, French horn, trumpet, double bass, flute, and trombone) has also been associated with poorer hearing (Axelsson and Lindgren 1981; Karlsson, Lundquist et al. 1983; Eaton and Gillis 2002). In many cases, this may also be related to the musician's position in the orchestra, with those sitting in front of the brasses or percussion section reporting worse hearing (Westmore and Eversden 1981). Violinists and flautists may also be at increased risk for developing worse hearing losses in their left and right ears, respectively, due to the placement of their instruments (Ostri, Eller et al. 1989).

Many researchers have noted that male musicians tend to have more pronounced hearing losses than females; whether this has a biological cause, or is a function of men playing louder instruments is not clear (Axelsson and Lindgren 1981; Steurer, Simak et al. 1998; Kahari, Axelsson et al. 2001; Kahari, Zachau et al. 2003). Musicians who have had a longer career are also at increased risk of hearing loss. In older people the natural loss of hearing with age (called 'presbycusis') seems to be worsened with increasing noise exposure (Kahari 2001). Since musicians and entertainment professionals often spend a great deal of their leisure time partaking in the music of others, it is important that they use hearing protection during these times to give their ears a rest (DeLay, Hiscock et al. 1991). If they do not, they increase their risk of hearing loss in the long run (Bray, Szymanski et al. 2004). Poor acoustic design of venues, especially important for rock music, may pose a greater risk for musicians' hearing as well (Hart, Geltman et al. 1987).

Given the substantial body of evidence there is ample reason to believe that musicians are at risk for hearing related issues, Chapter "Hearing health care for musicians" summarises measures which need to be taken up to prevent such risk.

REFERENCES

Arnold, G. & Miskolczy-Fodor, M. D. (1960).Pure-tone thresholds of professional pianists. *Archives of Otolaryngology, 71*, 938-947.

Axelsson, A. & Prasher, D. (2000). Tinnitus induced by occupational and leisure noise. *Noise and Health, 8,* 47-54.

Axelsson, A. and A. Ringdahl (1989)."Tinnitus - a study of its prevalence and characteristics. *British Journal of Audiology, 23*, 53-62.

Axelsson, A., & Lindgren, F. (1981).Hearing in classical musicians. *Acta Otolaryngol suppl 377, 3-75.*

Axelsson, A., & Lindgren, F. (1977). Does Pop Music Cause Hearing Damage? *International Journal of Audiology, 16*(5), 432-437.

Backus, B.C., Clark, T., & Williamon, A. (2007).Noise exposure and hearing thresholds among orchestral musicians. In A. Williamon & D. Coimbra (Eds.), Proceedings of the International Symposium on Performance Science, 23-28.

Behar, A., MacDonald, E., Lee, J., Cui, J., Kunov, H., & Wong, W. (2004).Noise exposure of music teachers. *Journal of Occupational and Environmental Hygiene, 1*(4), 243-247.

Behroozi, K. B. & Luz, J. (1997). Noise-related Ailments of Performing Musicians: A Review. *Medical Problems of Performing Artists, 12*(1), 19-22.

Bray, A., M. Szymanski, et al. (2004). Noise induced hearing loss in dance music disc jockeys and an examination of sound levels in nightclubs. *Journal of Laryngology And Otology, 118*(2), 123-128.

Burns, W. (1968).*Noise and man*. London: Murray.

Chasin, M. (2009).*Hearing loss in musicians: Prevention & management*. San Diego: Plural Pub.

Chesky, K. & Henoch, M. A. (2000). Instrument-specific reports of hearing loss: Differences between classical and nonclassical musicians. *Medical Problems Of Performing Artists,15*(1), 35-38.

Deatherage, P. (2003).*Effects of Music on Extended High Frequency Hearing*.(Electronic Thesis or Dissertation).Retrieved from https://etd.ohiolink.edu/

DeLay, S., S. Hiscock, et al. (1991). The effects of music technology on hearing: A case study of St. John's bars. *Canadian Acoustics, 19(*4): 77-78.

Drake-Lee, A.B. (1992). Beyond music: auditory temporary threshold shift in rock musicians after a heavy metal concert. *J R Soc Med, 85*(10): 617-619.

Early, K. L. & Horstman, S. W. (1996). Noise Exposure to Musicians during Practice. *Applied Occupational and Environmental Hygiene 11*(9), 1149-1153.

Early, K. L., & Horstman, S. W. (1996).Noise Exposure to Musicians during Practice. *Applied Occupational and Environmental Hygiene.*

Eaton, S. & Gillis, H. (2002).Review of orchestra musicians' hearing loss risks. *Canadian Acoustics 30*(2), 5.

Emmerich E., Rudel L. & Richter F. 2008. Is the audiologic status of professional musicians a reflection of the noise exposure in classical orchestral music? *Eur Arch Otorhinolaryngology, 265,* 753–758.

Fearn, R. W. & D. R. Hanson (1989). Hearing level of young subjects exposed to gunfire noise. *Journal of Sound and Vibration 131*(1), 157.

Fearn, R. W. (1976). Hearing-Loss Caused By Different Exposures to Amplified Pop Music. *Journal of Sound and Vibration 47*(3), 454-456.

Fearn, R. W. (1993). Hearing loss in musicians. *Journal of Sound and Vibration, 163*(2), 372.

Fisk, T., Cheesman, M. F., & Legassie, J. (1997). Sound pressure levels during amplified orchestra rehearsals and performances. *Canadian Acoustics 25*(3), 22.

Griffiths, S. K. &. Samaroo, A. L (1995).Hearing Sensitivity among Professional Pianists. *Medical Problems of Performing Artists, 10*(1), 11-17.

Gunderson, E., J. Moline, et al. (1997). Risks of Developing Noise-Induced Hearing Loss in Employees of Urban Music Clubs. American Journal of Industrial Medicine.

Hart, C.W., Geltman, C.L., Schupbach, J., Et Santucci, M. (1987). The musician and occupational sound hazards. *Medical Problems of Performing Artists, 2*(3), 22-25.

Hazards. *Medical Problems of Performing Artists, 2*(1), 22-25.

Henoch, M. A. & Chesley K. (2000). Sound exposure levels experienced by a college jazz band ensemble - Comparison with OSHA risk criteria. *Medical Problems Of*

Ising, H., W. Babisch, et al. (1997). Loud music and hearing risk. *Journal of Audiological Medicine, 6*(3), 123-133.

Jansen, E., Helleman, H., Dreschler, W., & de Laat, J. (2009). Noise induced hearing loss and other complaints among musicians of symphony orchestras. *International archives of occupational and environmental health 82*, 153-164.

Jansen, E., Helleman, H., Dreschler, W., & de Laat, J. (2009). Noise induced hearing loss and other complaints among musicians of symphony orchestras. *International archives of occupational and environmental health 82*,153-164.

Jansson, E., & Karlsson, K. (1983). Sound levels recorded within the symphony orchestra and risk criteria for hearing loss. *Scandinavian Audiology, 12,* 215.

Jaroszewski, A. & Rakowski, A. (1994). Loud music induced thresholds shifts and damage risk prediction. *Archives of Acoustics, 19*(3), 311-321.

Jaroszewski, A., Fidecki, T. et al. (1998). Hearing damage from exposure to music. *Archives of Acoustics 23*(1), 3-31.

Johnson D.W., Sherman R.E., Aldridge J. & Lorraine A. (1986).Extended high frequency hearing sensitivity: A normative study in musicians. *Annals of Oto-Rhino Laryngology, 95,* 196–202.

Johnson, D. W., Sherman, R. E. et al. (1985). Effects of instrument type and orchestral position on hearing sensitivity for 0.25 to 20 kHz in the orchestral musician. *Scandinavian Audiology, 14,* 215-221.

Juman, S., Karmody, C. S. et al. (2004).Hearing loss in steelband musicians. *Otolaryngology-Head and Neck Surgery 131*(4), 461-465.

Kähäri K., Zachau G., Eklöf M., Sandsjö L. & Möller C. 2003.Assessment of hearing and hearing disorders in rock/jazz musicians. *Int J Audiol, 42,* 279–288.

Kahari, K. R. (2001). Hearing assessment of classical orchestral musicians. *Scandinavian Audiology 30*(1), 13-23.

Kahari, K. R., Axelsson, A. et al. (2001). Hearing development in classical orchestral musicians. A follow-up study. *Scandinavian Audiology 30*(3), 141-149.

Kahari, K., G. Zachau, et al. (2003). Assessment of hearing and hearing disorders in rock/jazz musicians, *International Journal of Audiology 42*(5), 279-288.

Kahari, K., G. Zachau, et al. (2004). The influence of music and stress on musicians' hearing. *Journal of Sound and Vibration, 277*(3): 627.

Kahari, K.R., Axelsson, A., Hellstrom, P., Et Zachau, G. (2001). Hearing development in classical orchestral musicians. A follow-up study. *Scandinavian Audiology, 30*(1), 141-149.

Karlsson K., Lundquist P.G., Olaussen T. (1983). The hearing of symphony orchestra musicians, *Scandinavian Audiology, 12* (4), 257–264.

Karlsson, K., Lundquist, P. G. et al. (1983). The hearing of symphony orchestra musicians. *Scandinavian Audiology 12*(4), 257.

Kazkayasi, M., Yetiser, S. & Ozcelik, S. (2006). Effect of musical training on musical perception and hearing sensitivity: conventional and high frequency audiometric comparison. *The Journal of Otolaryngology, 35*(5), 343-348.

Laitinen, H. M., E. M. Toppila, et al. (2003). Sound exposure among the Finnish National Opera personnel. *Applied occupational and environmental hygiene 18*(3),

Lebo, C. P. & Oliphant, K. P. (1968).Music as a source of acoustic trauma. *Laryngoscope 78*, 1211-18.

Lee, J., A. Behar, et al. (2003). Noise exposure of opera orchestra players. *Canadian Acoustics, 31*(3), 78.

Lockwood, A. H., Salvi, R. J. et al. (2001). Tinnitus and the performer. *Medical Problems Of Performing Artists 16*(4), 133-135.

Maia J.R. & Russo, I.C. (2008).Study of the hearing of rock and roll musicians, *Pro Fono, 20*(1), 49–54.

McBride D., Gill F., Proops D., Harrington M., Gardiner K. et al. 1992.Noise and the classical musician. *British Medical Journal, 305*, 1561–1563.

McBride, D. (1992). Noise And The Classical Musician. *British Medical Journal305*(6868), 1561-1563.

Metternich, F. U. & Brusis, T. (1999). Acute hearing loss and tinnitus related to strongly amplified music. *Laryngo-Rhino-Otologie 78*(11), 614-619.

Meyer-Bisch, C. (1996). Epidemiologic evaluation of hearing damage related to strongly amplified music (personal cassette players, discotheques, rock concerts) - High definition audiometric survey on 1364 subjects. *Audiology, 35*, 121-142.

Mikl, K. (1995). Orchestral music: an assessment of risk. *Acoustics Australia, 23*(2), 51.

Miyakita, T., Hellstrom, P. A., Frimansson, E., & Axelsson, A. (1992).Effect of low level acoustic stimulation on temporary threshold shift in young humans, *Hearing Research, 60*, 149-155.

Nodar, R. H. (1993). Hearing Loss in a Professional Organist, A Case Study. *Medical Problems of Performing Artists 8*(1): 23-24.

Ostri, B., Eller, E., Dahlin, E., & Skylv, G. (1989).Hearing impairment in orchestral musicians. *Scandinavian Audiology 18*, 243-249.

Ostri, B., N, Eller, et al. (1989). Hearing impairment in orchestral musicians. *Scandinavian Audiology,18*(4), 243.

Palin, S. L. (1994). Does Classical-Music Damage The Hearing Of Musicians - A Review Of The Literature. *Occupational Medicine-Oxford 44*(3), 130-136.

Patel, J. A. and K. Broughton (2002). Assessment of the noise exposure of call centre operators. *Annals of Occupational Hygiene, 46*(8), 653-661.

Phillips, S. & Mace, S. (2008). Sound level measurements in music practice rooms. *Music Performance Research 2*: 36-47.

Rajalakshmi, K. and Apoorva, H.M (2012) Hearing in Musicians. Unpublished department project carried out at the department of Audiology, All India Institute of Speech and Hearing, Mysore.

Royster, J. D., Royster, L. H. et al. (1991).Sound exposures and hearing thresholds of symphony orchestra musicians. *Journal of the Acoustical Society of America, 89*(6), 2793.

Royster, J., Royster, L., & Killion, M. (1991).Sound exposures and hearing thresholds of symphony orchestra musicians. *Journal of the Acoustical Society of America 89* (6), 2793-2803.

Sabesky, I. J. & Korczynski, R. E. (1995).Noise Exposure of Symphony Orchestra Musicians. *Applied Occupational and Environmental Hygiene, 10*(2), 131-135.

Sataloff, R. T. (1991). Hearing Loss in Musicians. *American Journal of Otology, 12*(2), 122-127.

Sataloff, R. T. (1997). Hearing loss in singers and other musicians. *Medical Problems Of Performing Artists, 12*(2), 51-56.

Schmidt, J. M., Verschuure, J. et al. (1994). Hearing-Loss In Students At A Conservatory. *Audiology, 33*(4), 185-194.

Schmuziger, N. J., Patscheke, J., & Probst, R. (2006).Hearing in non-professional pop/rock musicians. *Ear and Hearing , 27*, 321–330.

Smeatham, D. (2002). Noise levels and noise exposure of workers in pubs and clubs - a review of the literature. Research Report 026: 91.

Spoor, A., & Passchier-Vermeer, W. (1969). Spread in Hearing-Levels of Non-Noise Exposed People at Various Ages. *International Journal of Audiology.*

Steurer, M., Simak, S. et al. (1998). Does choir singing cause noise-induced hearing loss? *Audiology, 37*(1): 38-51.

Szymanski, K. (1983). The sound of music noise levels in concerts. *Noise & Vibration Control Worldwide, 14*(2), 54.

Teie, P. U. (1998). Noise-induced hearing loss and symphony orchestra musicians: risk factors, effects, and management. *Maryland Medical Journal 47*(1), 13-18.

Westmore G. A. & Eversden I. D. (1981) Noise-induced hearing loss and orchestral musicians. *Arch Otolaryngol Head Neck Surg.107* (12), 761-764.

Chapter 7

HEARING HEALTHCARE FOR MUSICIANS

"Without music, life is a journey through a desert."

Pat Conroy

Hearing empowers and allows us to work, socialize, interact, communicate and relax. Hearing is required for safety and to keep alert to the world around us. In this modern age of computers, telephones and mass media, one of the basic ways in which we receive information is through our sense of hearing. Whether we are at work engaged in a meeting, listening to music or listening to our partner share their thoughts over dinner, accurate hearing is vital.

The human ear is divided into three parts: the outer, middle, and the inner ear. The outer ear has the pinna and ear canal which leads to the ear drum; the middle ear has three tiny ossicles and two windows, the inner ear has a snail shaped organ called cochlea which is connected to the brain through the auditory nerve. The length of the auditory nerve is about two centimeters. You will be amazed to find that this two centimeter nerve transmits all that we hear to the brain. Hearing loss due to outer and middle ear is termed as conductive; inner ear (cochlea) losses are known as sensorineural.

There are various causes for developing hearing loss and exposure to loud sounds is one amongst them. Excessive sound levels results in damage of the inner ear. There are tiny hair cells inside the inner ear which vibrate with sound. When these hairs are exposed to loud noise, even a short loud burst, the tip of the hair cells send false information to the brain. Sometimes these tips of the hairs can break off; causing a ringing sensation that can take some time to go away. The tips grow back and this is why the effect is temporary. If one is exposed to high sound levels, then whole clumps of cells can be broken off - and these do *not* grow back, causing permanent hearing damage.

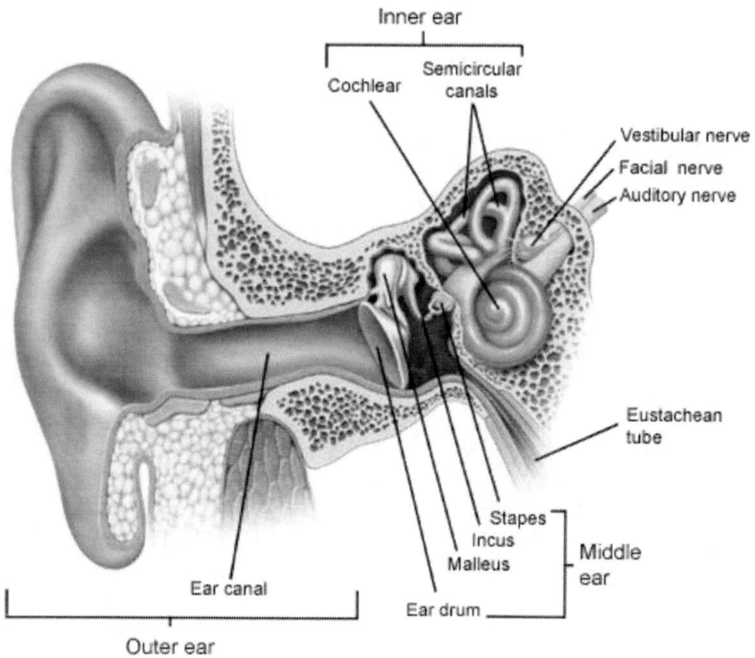

Figure 7.1.Major parts of the ear.

Music is most frequently wanted sound (unlike noise), listened to with pleasure and cannot be compared to industrial noise (unwanted sound) causing permanent Noise Induced Hearing Loss (NIHL). However, since music also can be characterized by its loudness, duration and quality, it may induce damage to the inner ear resulting in hearing impairment to a minor or major extent depending on the individual susceptibility to noise (Burns, 1968). Results of several studies (Jansen et al, 2008; Emeriti et al, 2007; Kothari et al, 2004) indicate the prevalence of hearing loss in professional musicians to be 38–50%. Investigators (Behar, MacDonald, Lee, Cui, Kunov & Wong, 2004) have reported that 78% of the music teachers were exposed to sound levels exceeding 85 dB (A), when measured with a noise dosimeter. The musical professionals are at risk of developing hearing loss as they are exposed to high levels of sounds. For musicians, the sense of hearing can be considered as the most important 'instrument' as they are highly dependent upon a good sense of hearing; both with respect to hearing very faintly presented tones and the ability to discriminate between different frequencies. Consequently, this 'instrument' should be treated with utmost care, to protect the auditory sense in order to maintain musical perception and ability. Below are a few preventive measures which need to be followed during practice and performance.

GENERAL DO'S AND DON'TS TO PREVENT HEARING LOSS

Do's

- ***Be a Quiet Enforcer***

Turn down the ambient noise level in your life by buying appliances and devices that have low noise ratings.

- ***Wear Hearing Protection***

Wear ear protection devices like ear plugs if you know you're going to be exposed to loud sounds including loud music for more than a few minutes. Musicians' earplugs are designed to attenuate sounds evenly across the frequency range and thus minimize their effect on the user's perception of bass and/or treble levels. Musicians' earplugs are a custom product, made individually for each user and can be purchased only from licensed hearing professionals.

- ***Ear Wax Removal***

Wax buildup in your ears can muffle sound. Usually through a natural cleansing mechanism ear wax gets out of the ears. If wax has become compacted in your ear, your doctor may need to remove it. But don't use a cotton swab to clean them out. Cotton swabs can push wax even deeper into your ear canal.

- ***Check Medications Before Consumption***

Some 200 medications are potentially ototoxic, or damaging to hearing. Ototoxic medications include some antibiotics and certain cancer-fighting drugs, among others. Even high doses of aspirin can harm your ears. If you take a prescription medication, check with your doctor to make sure it doesn't pose a threat to your hearing. If you must take a medication that may be ototoxic, make sure your doctor monitors your hearing and balance before and during your treatment.

- ***Have Your Hearing Tested Routinely***

Make an appointment to get your hearing tested if you:

- Have close relatives with hearing loss
- Have trouble hearing conversations in noisy or even in silent environments
- Are exposed to loud noises on a regular basis
- Experience a frequent ringing or buzzing sound in your ears
- Ear pain or discharge from ears

Don'ts

- ***Don't Expose to Excessive Noise***

Exposing your ears to high level noise, such as gunfire, damages the inner ear even if the duration is just 15 minutes. The end result could be permanent hearing loss and/or ringing in the ears (tinnitus). Earplugs commonly known as "shooters protection" restrict the amount of harmful frequencies and loud noise from damaging the ear.

- ***Do Not Listen with Ear Phones/Headphones with High Volume***

Listening to music through headphones at a high volume for extended periods of time can result in permanent hearing damage.

- ***Don't Smoke and Avoid Alcohol***

Exposure to tobacco smoke has been linked to increased risk of hearing loss. Research has shown that smoking, age, and noise exposure can collectively increase a person's risk for hearing loss. If you smoke, preserving your hearing is one more good reason to quit. If you don't smoke, be sure to avoid breathing second hand smoke.

- ***Don't Insert Sharp Objects into Ears***

Ear drum, a thin membrane is located at the end of our ear canal. Inserting sharp objects into your ears may damage the ear drum and lead to hearing problems.

- ***Don't Self Medicate***

Self medication can be injurious. The medicine you take without doctor's prescription may be harmful to the ear and might lead to hearing loss.

Hearing is crucial for musicians to continue performing at a standard, which is expected, and accustomed. Therefore monitoring the effects of loud music on musicians' hearing is imperative.

HEARING PROTECTION

If one is exposed to high sound levels, then whole clumps of cells can be broken off - and these do *not* grow back, causing permanent hearing damage. Musicians are a group of individuals who are exposed to higher levels of sound during their practice sessions as well as concerts. This necessitates proper hearing protection for them in order to avoid permanent hearing damage.

Proper hearing protection can be achieved by the use of hearing protection devices. For e.g. standard ear plugs are used by several musicians to avoid exposure to high level sounds. These ear plugs are available in standard sizes. However, custom made hearing protection can be obtained at an audiology or hearing aid dispensing facility where custom made impressions are taken to process and prepare custom ear plugs. However, many musicians complain that when hearing protection is tried it renders the sound hollow and there is "no high-end". This is experienced due to a phenomenon called occlusion effect. The occlusion effect is a build-up of low-frequency sound energy in the ear canal from a person's own voice or musical instrument. When an ear is plugged up, a person may report that sound is "hollow" or "echoey". It happens because the sound gets trapped in the ear canal and goes inward through our hearing system.

Understanding the reasons for this and physics behind this, many improved and readily acceptable ear plugs are available today to overcome these problems. Musicians today can obtain optimal hearing protection with no echoing of sounds. In order to minimize the occlusion effect, a longer earplug is made that extends in to the boney portion of the ear canal. If the ear plug is made with a long enough bore extending into the ear canal, it will prevent any vibration from being transduced to the ear. Practically this means that in order to minimize the occlusion effect, one need to have a custom ear plug that is made with a long ear canal bore. Another solution to minimize the occlusion effect is to create a hole or vent in the ear plug that will "bleed off" the unwanted low-frequency sound energy. There are ear plugs which were developed especially for musicians which ensure that there is equal attenuation at all frequencies and the discrepancy between low and high frequencies are eliminated. They are listed below.

The ER-15 Earplug

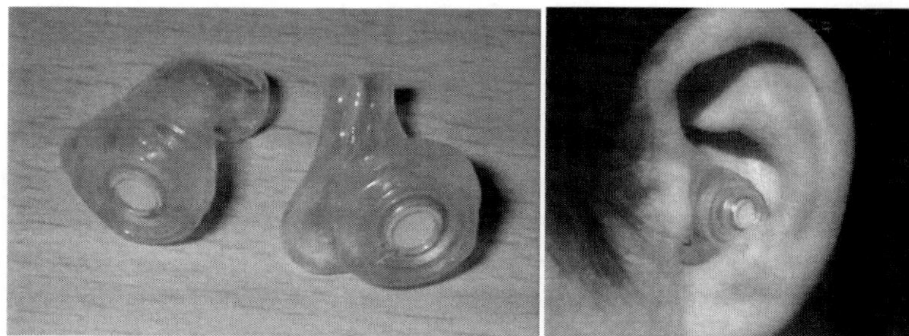

Figure 7.2. A. Picture of ER-15 ear plug. B. Ear plug being worn.

In 1988, a manufacturer named Etymotic Research, Inc. (hence the name ER) devised a custom earplug with approximately 15 decibels of attenuation over a wide range of frequencies. The earplug (named the ER-15) has become widely accepted by musicians as well as some industrial workers who work in relatively quiet environments. That is, the ER-15 makes all the sounds of music equally quieter by about 15 dB.

The Vented/Tuned Earplug

A vented/tuned earplug is a custom earplug with a tuned hole or cavity drilled down the center. The diameter of the hole can be adjusted (tuned) with covers (called select-a-vents [SAV]) that have various sized holes in them. In its most open position, the vented/ tuned earplug is acoustically transparent below 2000 Hz. Most of the musicians who wear this form of earplug use it in the most open position. A vented/tuned earplug would allow a musician to hear the lower and mid-frequency sound energy of their own instrument but provide significant attenuation for the high frequency stimuli in the immediate vicinity. Such an earplug is quite useful for woodwinds or larger stringed instruments that have significant high-frequency energy in their environments which could be damaging to their ears. Ear mold laboratories have different names for these "filtered" vented/ tuned earplugs, and the attenuation characteristics will vary depending on the filter used, as well as on the sound bore dimensions.

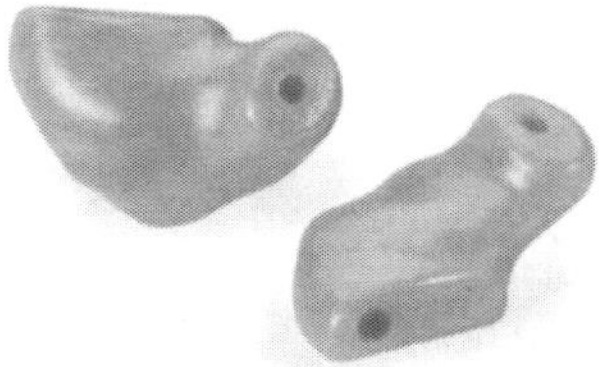

Figure 7.3. A Picture of a Vented ear plug.

Instrument Specific Strategies

Different musicians find themselves in differing environments, and hence the strategies to reduce the effects of music exposure may be different. Some of the strategies for various instruments are listed below:

1. Small Stringed Instrument
The violins and violas belonging to this category, generate significant low and high-frequency sounds and more importantly, these musicians need to hear the balance between these sounds. The ear protection of choice is the ER-15 earplug, as this will provide broadband uniform attenuation that is ideal for these musicians.

Figure 7.4. A Picture of a string instrument:Violin.

A very important factor is the positioning of the musician. Violin and viola players should never be placed under an overhang that is within 1 meter of their heads.

2. Large Stringed Instruments

Figure 7.5. A Picture of a large string instrument: Harps.

The cello, bass, and harp are the three major instruments in this category. Like all stringed instruments, it is crucial to hear the high-frequency harmonic content; however, because of the larger physical size, most of the important harmonic energy is below 2000 Hz (one octave below the top note on a piano keyboard). Therefore, unlike violin and viola players, placing these musicians under a poorly constructed performance pit overhang will subsequently have little acoustic and ergonomic effect on their playing. Partly because of the large size of these instruments, the sound levels generated are not excessively great. However, the brass section is typically located to their immediate rear, and it is this source that is the potential threat to hearing loss. The ear protection of choice is the vented/tuned earplug. This earplug serves to allow most of the fundamental and higher frequency harmonic energy of these instruments through to the ear with minimal alteration in the spectrum.

3. Woodwinds

Figure 7.6. A Picture of a wind instrument:Flute.

The woodwind category is made up of clarinets, saxophones, flutes, oboes, and bassoons, as the major representatives. These instruments can be played in a wide range of venues, such as in large orchestras or with small jazz and blues bands. In orchestral environments, woodwinds are typically near the front of the orchestra but situated in front of the brass or percussion sections. In jazz and blues environments, they can be in front of speakers or near the drummer. The *ER-15 earplug* is recommended for jazz and blues performing venues, and *vented/tuned earplugs* for the orchestral environments. The rationale for the use of the vented/tuned earplugs in an orchestra is similar to that for the large stringed instruments—namely, to attenuate the high-frequency music energy from other instruments to the rear. The broader band ER-15 earplugs are more appropriate for the more varied music sources in a jazz or blues band.

With woodwind instruments, the sound emanates from the first noncovered finger hole, and as such may be "lost" between the legs and music stands of other musicians. Placing woodwind players in an unobstructed location will allow a direct path for their sound. If the woodwinds' sound is obstructed, overplaying can and does result with the potential of a career-ending injury to the muscle that controls the embouchure. This is also a frequently stated concern of brass musicians. In the jazz and blues environment, a woodwind player may be situated near a speaker or near the cymbals of the drum kit. In most jazz or blues bands, the drummer hits harder on the ride cymbal (on the dominant side—typically the right side) and the woodwind player should situate himself or herself away from that potential source of music exposure. It is usually the best strategy to situate oneself parallel to a speaker rather than to the front or to the rear of the speaker, as the speaker enclosure wall affords some protection.

4. Percussion Instruments

In terms of ear protection unlike other musicians, there is very little choice for most percussionists. The ER-25 earplug is the optimal ear protection. Excessive ear protection may result in wrist or arm damage, and too little may result in a progressive music-induced hearing loss. For the full drum kit (with drums and cymbals), hitting the high hat cymbal with a drum stick can be the greatest potential threat to hearing.

Figure 7.7. A Picture of a percussion instrument:Drum set.

It is commonplace to find that the hearing in the left ear is worse for a right handed drummer because of the close proximity to the high hat (and the converse for some left handed drummers). The high hat is the main cymbal for rock and roll, with the ride cymbal (on the dominant or, typically, right side) for jazz and blues. The high hat, as the name suggests, is about the diameter of a brimmed hat with two opposing cymbals facing each other. A pedal can be used to move them together or leave them apart. Rock and roll musicians tend to play with the high hats slightly apart, whereas blues and jazz musicians tend to have them apart only about 50% of the time. There will be increased spectral bandwidth and increased intensity when the high hats are left open and the duration of the sound of the open high hat will be longer. The positioning of the high hats is a matter of taste (and music style), and little can be done to convince a drummer to change his or her taste other than by education. However, many drummers are willing to play with a closed position high hat during practice sessions or even to use a muffling practice pad in between the two opposing faces of the high hat cymbal.

When played in the closed condition, the overall intensity is less. Another innovation that derives from the military, has recently been resurrected for drummers (and electric bass players) called seat shakers (or bass shakers). These very low frequency woofer loudspeakers can be hooked in to the amplification system and will generate improved low frequency vibrational feedback. During playing, low-frequency vibrational feedback is available to the musician, providing him or her with better monitoring. Drummers using this device, tend not to play as loud with a subsequent reduction in the level of the music. This potentially decreases the risk of arm and wrist injury.

As an alternative to earplugs, personal earphones that replace stage monitors, can also be an option. These earphones (or in-the-ear monitors) can be custom or noncustom. They are essentially in-the-ear hearing aid shells that use a broadband speaker that can handle high-levels with minimal distortion. These are either hard wired to the amplification system or connected to an FM wireless receiver worn on the belt or under the clothes. The music is mixed by the sound engineer and transmitted back to the musician so that there will be optimal monitoring. In the wireless situations, the musician is free to move about the stage. Because the effects of the environment are minimized when these earphones are used, the sound levels at the ear are significantly lower than if the musicians were using conventional sound field monitoring systems such as wedge speakers.

A less costly "hard-wired" system can be used by joining a pair of broad band hearing aid receivers to standard (body aid style) earmolds and connecting this to an equalizer/amplifier that can be adjusted for overall gain and frequency response shaping. This has been very useful for percussionists who only need improved monitoring to hear a "click track". Some recent research has indicated that people still need to be aware to turn down the volume when they wear in-the-ear monitors. It turns out that quieter, less damaging levels are more acceptable by those who wear the monitors (versus stage monitors) but that unless counselled, people may still wear them at higher potentially damaging levels.

Finally, drummers can use the effect of their stapedial reflex to their advantage. If a loud sound is about to occur, such as by hitting a cymbal, the drummer can begin humming just prior to and through the length of the loud sound. Such humming serves to elicit the stapedial reflex, thus providing some additional temporary ear protection.

5. Amplified Instruments

Musicians in amplified environments usually have a more flexible work environment than those who work in an orchestra. Unlike many orchestral musicians, they may have some control over the instrument set-up. Therefore, recommendations of an environmental change for reducing the potential for music exposure are more frequently followed.

These musicians should be either situated away from loud speakers or parallel to the speaker enclosures walls as this will afford some protection. In those cases where loud speakers are oriented towards the musicians in order to obtain "side-wash," the speakers should be elevated. This should be done with caution, as many loud speakers are designed to be placed on the floor. One should check with the manufacturer or retailer before using a speaker in a position that it was not originally intended. Elevating the speakers is done for the same reason as elevating brass players on risers. Like the trumpet, the speaker cone of a loud speaker is highly directional, but only for the higher frequency treble notes. Increasing the height of loud speakers to the level of the performer's ear will improve the monitoring of these higher frequency components, such that the sound engineer will be able to decrease the overall sound level emanating from the loud speaker. The musician will feel that the music is as loud as before, but the overall intensity will be lower, with less potential for hearing damage. Musicians playing in an amplified environment should be using the ER-15 earplug (except for the drummer who should be using the ER-25). If improved monitoring is required, in-the-ear monitoring (as discussed in the percussion section above) can be very useful. Other than improved monitoring, significant hearing protection can also be achieved with these monitors.

6. Vocalists

By far, the primary concern of vocalists is related to vocal strain. Blues and jazz singers frequently complain of sore throats after long sets, especially in smoky environments. While part of the problem is the air quality, their situation can be improved by the use of an earplug that creates a slight occlusion effect. Vocalists are divided into two categories: solo and nonsolo. Solo vocalists can be found in classical environments and may be accompanied by relatively quiet instruments. The *vented/ tuned earplug* is the most appropriate protection in this case. These earplugs will provide some high-frequency attenuation, but also will generate a slight increase in the loudness of one's own voice. The vocalist will tend to sing slightly less loud with reduced long-term vocal strain. Nonsolo vocalists, typically found in pop bands, are susceptible to the same sources of music exposure than their instrumental colleagues. Because of the wide range of music sources on stage, the *wide band ER-15 earplug* is the form of ear protection of choice. As in the case of the solo performer, a slight degree of occlusion may be useful, and it is therefore recommended that the ER-15 earplugs be made with short canals. Nonsolo vocalists will also gain an improved monitoring (as well as some hearing protection from the various sources of music) by using the custom earphones.

Healthy hair cells are the key to good hearing. Although, some die off naturally as you age, many more are killed early, from unprotected exposure to noise or loud music. The musicians will derive following advantages associated with the use of hearing protection: i) Conserve hearing by reducing the damage risk ii) Reduce or block out loud music generated by other musicians iii) Reduce the incidence of pain in the ears, temporary deafness, and ringing in the ears and iv) Improve sound quality.

Thus it becomes imperative to all the performing musicians to take of their hearing because "Prevention is better than cure" goes the wise saying.

REFERENCES

Control the volume (n.d.). Retrieved October 20, 2014, from http://uk.earpeace.com/

Musicians ear plugs (n.d.). Retrieved August 14, 2014, from http://www.etymotic.com/hp/erme.html

Chasin, M. & Chong, J. (1991).An in situ ear protection program for musicians. *Hearing Instruments, 42*(12), 26-28.

Chasin, M. & Chong, J. (1992).A Clinically Efficient Hearing Protection Program for Musicians. *Medical Problems of Performing Artists 7*(2), 40-43.

Chasin, M. & Chong, J. (1995).Four Environmental Techniques to Reduce the Effect of Music Exposure on Hearing. *Medical Problems of Performing Artists 10*(2), 66- 69.

Groothoff, B. (1999). Incorporating effective noise control in music entertainment venues? Yes, it can be done. *Journal of Occupational Health and* Safety (Aust. NZ), *15*(6), 543-550.

Hall, J. W. I. & Santucci, M. (1995).Protecting the professional ear: conservation strategies and devices. *Hearing Journal, 48*(3): 37-45.

Killion, M. C. & Stewart, J. (1988).An earplug with uniform 15-dB attenuation. *Hearing Journal 41*(5), 14-17.

Santucci, M. (1990). Musicians Can Protect Their Hearing. *Medical Problems of Performing Artists, 5*(4), 136-138.

Niu, X., & Canlon, B. (2002).Protecting against noise trauma by sound conditioning. *Journal of Sound and Vibration.*

AUTHOR'S CONTACT INFORMATION

Professor K. Rajalakshmi,
Department of Audiology
All India Institute of Speech & Hearing
Manasagangothri, Mysore 570 006
Karnataka, India
Email: veenasrijaya@gmail.com

INDEX

Y